CRANE
SAFETY

A Guide to OSHA Compliance
and Injury Protection

Carl O. Morgan, CSP, PhD

Government Institutes
Rockville, Maryland

Government Institutes, a Division of ABS Group Inc.
4 Research Place, Rockville, Maryland 20850, USA
Phone: (301) 921-2300
Fax: (301) 921-0373
Email: giinfo@govinst.com
Internet: http://www.govinst.com

03 02 01 00 99   5   4   3   2   1

Library of Congress Cataloging-in-Publication Data

Morgan, Carl O., 1943-
    Crane Safety: a guide to OSHA compliance and injury protection / Carl O. Morgan.
        p. cm.
    Includes bibliographical references.
    ISBN 0-86587-687-8 (pbk.)
    1. Cranes, derricks etc.--Safety measures. 2. Industrial safety--Law and legislation--United States. I. Title.

TJ1363 .M57 2000
621.8`73`0289--dc21

                                                          99-054937

ISBN: 0-86587-687-8

This book is dedicated to safety advocates
striving to make a safer place of employment
for all employees.

Summary Contents

Contents

LIST OF FIGURES AND TABLES

PREFACE

This book identifies the requirements necessary for the safe operation of cranes in construction and general industries while assuring compliance with the myriad of imposed laws and regulations.

A unique perspective is presented in this book by combining all the current rules and regulations on crane safety and presenting them in a single publication. Because no single source currently exists for all the information needed for the safe operation of cranes, this book fills a void for a very large but undetermined number of crane operators, and for other workers in industry.

The following features make this book unique:

- All the information needed for a crane safety program (overhead, gantry, mobile cranes, and derricks) is combined in this one publication. This includes rules and regulations from government agencies and crane manufacturers along with information about applicable American National Standards Institute (ANSI) standards.

- The OSHA standards included in the text are:

    29 CFR 1910.179 – Overhead and gantry cranes
    29 CFR 1910.180 – Crawler locomotive and truck cranes
    29 CFR 1910.181 – Derricks
    29 CFR 1910.184 – Slings
    29 CFR 1926.550 – Cranes and derricks
    29 CFR 1926.251 – Rigging equipment for material handling.

- Charts, checklists, and forms are provided to assist in creating a safe working environment when using overhead, gantry, mobile cranes, and derricks. An example of one form included in the appendices is a lift plan that can be used to meet the OSHA requirement of documenting in writing before the lifting of any personnel is conducted. (Appendix L)

- In addition to the index, definitions from pertinent OSHA standards and commonly used crane terminology are included in a glossary at the end of the text.

ABOUT THE AUTHOR

Carl Morgan is a Certified Safety Professional (CSP) with extensive experience in safety compliance in general industry and construction. While working at the Aluminum Company of America (ALCOA) in construction and safety, he obtained a bachelor of science (B.S.) degree with a concentration in industrial training and a master of public health (MPH) degree with a concentration in environmental, safety, and health from the University of Tennessee. Following ALCOA, Mr. Morgan went to work at the Oak Ridge National Laboratory (ORNL) in Oak Ridge, Tennessee, where he was the hoisting and rigging program manager for an extended period of time. While employed at ORNL, he earned a doctorate of philosophy degree (Ph.D.) at the University of Tennessee with a concentration in safety management. Dr. Morgan is a professional member of the American Society of Safety Engineers (ASSE) and is past president of the ASSE East Tennessee Chapter.

ACKNOWLEDGMENTS

Permission has been received from the Wire Rope Technical Board to reproduce or quote from their Wire Rope Sling User's Manual. The Crosby Group, Inc., granted permission to use information from the Crosby General Catalog and has provided the Crosby® User's Lifting Guide for inclusion in this text. The Yale Cordage Company also granted permission to use portions of the Yale Cordage Industrial Catalog. The support of these companies has made it possible to present this manual in its effective and comprehensive form.

1

MOTIVATION FOR SAFETY

As the number of cranes in operation in construction and general industry continues to increase, so do the number of employees connected with the various complex hoisting and rigging tasks. From the crane operator, riggers, supervisors, and all other personnel involved in the daily use of cranes, the motivation for crane safety has to become an inherent part of everyday activities to prevent daily occurrences causing personnel injury and property damage.

Because the potentiality exists for a serious accident in hoisting and rigging tasks that could result in consequential property damage, personnel injury, or indeed death, there is a compelling need for employers to be compliant with all the current crane-related rules and regulations. This includes defining the scope of the work, making sure all employees are trained and qualified to perform their assigned tasks, and using equipment maintained according to the manufacturer's requirements.

CRANE AND HOIST SAFETY

More than 250,000 crane operators,[1] a very large but undetermined number of other workers, and the general public are at risk of serious and often fatal injury due to accidents involving cranes, derricks, hoists, and hoisting accessories. There are approximately 125,000 cranes in operation today in the construction industry as well as an additional 80,000 – 100,000 in general and maritime industries.

According to the Bureau of Labor Statistics' (BLS) Census of Fatal Occupational Injuries, 79 fatal occupational injuries were related to cranes, derricks, hoists, and hoisting accessories in 1993. [2] OSHA's analyses of crane accidents in general industry and construction identified an average of 71 fatalities each year. [3] While adequate worker exposure data is lacking to calculate the risk of death for the entire population exposed, the risk of death among crane operators alone is significant, corresponding to a risk of more than one death per thousand workers over a working lifetime of 45 years. OSHA is developing an action plan to gather additional information and reduce worker exposures to this hazard, but is not initiating rulemaking at this time.

Hazard Description

In 1992, OSHA reviewed the accident investigation files of 400 crane incidents in general industry and construction over a five-year period and identified 354 fatalities—an average of 71 fatalities per year. [3] According to the 1987 BLS supplementary data system (23 states reporting), more than 1,000 construction injuries were reported to involve cranes and hoisting equipment. However, underreporting of crane-related injuries and fatalities due to misclassification and a host of other factors mask the true magnitude of the problem.

A 1989 catastrophic tower crane collapse in downtown San Francisco and a 1993 mobile crane accident near Las Vegas heightened public awareness to the continuing problem of crane accidents. Since crane activities normally occur in urban areas, unsafe equipment and operations present a risk not only to workers, but to the general public as well. Five citizens were killed in the incidents mentioned above; two in San Francisco and three in Las Vegas.

OSHA's analysis also identified the major causes of crane accidents to include: boom or crane contact with energized power lines (nearly 45% of the cases), under-the-hook lifting devices, overturned cranes, dropped loads, boom collapse, crushing by the counterweight, outrigger use, falls, and rigging failures. [3]

Some cranes are not maintained properly or are not inspected regularly to ensure safe operation. Crane operators often do not have the necessary qualifications to operate each piece of equipment safely, and the operators' qualifications required in existing regulations might not provide adequate guidance to employers. The issues of crane inspection/certification and crane operator qualifications and certification needs to be further examined.

Current Status of Crane Standards

OSHA's crane standards for construction, general industry, and maritime have not been updated since 1971 and rely heavily on outdated 1968 ANSI consensus standards. The current OSHA standards do not address many of the advancements in hoisting technology or equipment used in construction today, such as the climbing tower crane that failed in San Francisco.

In a study of occupational fatalities between 1980 and 1992, machinery-related incidents ranked as the fourth leading cause of traumatic occupational fatalities in the U.S. construction industry, resulting in 1,901 deaths and 2.13 deaths per 100,000 workers. [4] Most of the victims were males, having nearly eight times the fatality rate observed among females (2.3 vs. 0.29). The fatality rate in the Northeast census region, 1.29, was considerably lower than the rates in other regions, which ranged from 2.05 to 2.37. Overall, fatality rates declined 50% over the study period.

Workers in three occupation divisions—precision production, craft, and repair; transportation and material moving; and handlers, equipment cleaners, helpers, and laborers—had both the highest frequency and highest rate of fatalities. Cranes, excavating machinery, and tractors were the machines most frequently involved in fatalities in this study. During the study period, fatality rates for tractors and cranes declined 71% and 67%, respectively, while rates for excavating machinery declined only 12%. The most common crane incident types were; struck by a mobile machine, crane overturned, and struck by a boom.

Further delineation of groups at highest risk for machinery-related injuries is complicated by a lack of data on exposure to machinery since exposure is clearly not equivalent across all occupational groups within the construction industry. The findings suggest that injury prevention programs should focus not only on machine operators, but also on those who work around machines. Translation of regulations into "plain English," providing incentives for safe work practices, and addressing safety in project planning stages can also help to reduce machinery-related deaths in the construction industry.

On October 19, 1992, OSHA issued an Advance Notice of Proposed Rulemaking and requests for comments on crane safety in general industry and construction (Federal Register # 57: 4774). OSHA was considering a multi-phased revision of the crane safety provisions of 29 CFR Part 1926, Subpart N (Cranes, Derricks, Hoists, Elevators, and Conveyors) and of 29 CFR 1910, Subpart N, (Materials Handling and Storage). One of the primary areas of concern to the agency was the limited criterion for crane operator qualifications incorporated by reference into the existing regulations.

Other areas OSHA announced that they would explore and evaluate included:

- The need to update parts of the standard dealing with the use, inspection, and maintenance of cranes

- The need for a requirement for certification of cranes that are used on construction sites and general industry sites

- The need for a requirement for certification of riggers and signal persons.

Written comments on the advance notice of proposed rulemaking were to be postmarked by February 12, 1993. However, on March 31, 1993, the notice was withdrawn with no action taken.

In 1994 the National Advisory Committee on Occupational Safety and Health recommended that crane safety be given a high priority (11/30/94). Also in 1994, the National Institute for Occupational Safety and Health (NIOSH) completed the revision of a previous NIOSH alert on crane-related electrocutions based on a review of recent data from the National Traumatic Occupational Fatalities (NTOF) surveillance system and recommendations from the Fatality Assessment and Control Evaluation (FACE) program. This alert, entitled "Preventing Electrocutions of Crane Operators and Crew Members Working Near Overhead Power Lines," was published in May 1995. [5]

The American Society of Mechanical Engineers (ASME) and ANSI, through consensus procedures, continually update standards for crane manufacturing, operational procedures, inspection requirements, and operator qualifications (B30 series, 1986-1994). [6] Many construction employers adopt these standards in order to maintain their equipment and reduce losses, both in human terms and property damage. In addition, the Specialized Carriers and Riggers Association (SC&RA), which represents most of the construction crane owners and users, has developed a set of requirements for crane operator qualification and certification.

Rationale

Crane and Hoist Safety meets several of the criteria for designation as an OSHA priority:

- The very serious nature of the hazard

- The magnitude of the risk (a high rate of fatalities and serious injuries relative to the number of workers exposed)

- The potential for catastrophic accidents

- The considerable knowledge available about effective protective measures.

All of the above clearly demonstrates the need for action to address crane and hoist safety.

Since crane activities normally occur in metropolitan areas, unsafe equipment and operations present a risk not only to workers, but to the general public as well. The catastrophic tower crane collapse in 1989 and the 1993 mobile crane accident mentioned earlier heightened public awareness to the continuing problem of crane accidents.

Crane Incidents at the Department of Energy

To a crane operator, few experiences can be as frightening as when a crane becomes unbalanced while a load is being lifted or when the crane collapses under the weight of an excessive load. In April 1993, cranes became unbalanced during two separate incidents at Department of Energy (DOE) sites. The following discusses these incidents, as well as a related industrial incident in which a crane collapsed.

The first of the two incidents at DOE sites occurred on April 28, 1993, when a crane at the Hanford Site became unbalanced while the boom was being lowered. The operator was preparing to lift a waste cask and made the mistake of lowering the boom too far, thereby shifting the crane's center of gravity. As the crane's rear outriggers lifted off the ground, the operator quickly raised the boom and restored the crane's balance. No damage occurred, and no one was injured.

The primary cause of the incident was attributed to the operator's lack of familiarity with the crane. This particular crane had five outriggers rather than four, a more common configuration. The operator was unaware of the fifth outrigger and had not extended it to stabilize the crane. Using the fifth outrigger could have prevented the incident.

The second DOE incident occurred two days later, on April 30, 1993. A worker at the Bryan Mound site used an 18-ton mobile crane to off-load an 8,300-pound electric motor. The worker lifted the motor and then rotated the boom to position and lower the motor. The weight of the load caused the crane to tip forward. The worker quickly lowered the load, preventing any damage to the crane. However, the electric motor being lifted and two others located on the ground were damaged.

Although the worker in this instance was a certified crane operator, his lack of experience contributed to the incident. He underestimated the weight of the lift by 2,000 pounds. After lifting and rotating the load successfully, he attempted to lower the boom in order to extend the reach of the crane to the drop zone, at which point the crane became unstable.

The Industrial Incident

Failure to determine the correct weight of a load had more serious consequences in an industrial accident that occurred in Dayton, Ohio on March 28, 1991. In this incident, a crane collapsed while removing an air-conditioning unit from the roof of a building. The crane operator had been told that the unit weighed 3,400 pounds; it actually weighed 5,510 pounds. The crane had a rated capacity of only 4,800 pounds.

OSHA has cited the operator of the crane for failing to take reasonable steps to verify the weight of the load, a violation of 29 CFR 1926.500(a)(1). Because the crane operator was a subcontractor to the rigging company hired to remove the air-conditioning unit, subcontractor management argued that the rigger was responsible for determining the unit's weight.

This argument did not sway the Secretary of Labor, who maintained that the crane operator should have taken reasonable steps to verify the information provided by the rigging company. According to the brief filed recently in the Sixth Circuit of the U.S. Court of Appeals, no one from the subcontractor organization examined the unit before the lifting operation began. The unit's size should have led workers to question the weight estimate. The crane operator had a second opportunity to question the 3,400-pound estimate when it was learned that the replacement unit weighed 5,000 pounds. Finally, even if the 3,400-pound estimate had been correct, the additional weight of crane attachments meant that the total load weight would have come within 100 pounds of the crane's rated capacity.

Lessons Learned From These Incidents

These incidents suggest several general guidelines for the safe use of cranes:

- Crane operators must know the weight of a load. If a crane operator is unsure about the weight of a load, he or she should refer to the shipping ticket or other documentation. Crane operators should also ensure that all calculations regarding lifts are correct; for example, the hook distance should be measured from the crane's centerline of rotation, not from the base of the boom.

- Crane operators must be trained and qualified to operate their equipment. Completion of a training course does not ensure that a worker is qualified; on-the-job training is an important supplement to formal instruction. Training isn't complete until proficiency can be consistently demonstrated.

- Qualified personnel should supervise crane operations. Appropriate supervision is another important supplement to formal instruction. The inexperienced crane operator at Bryan Mound was not supervised during the lift.

- Crane operators must be familiar with their equipment. The operator associated with the Hanford incident was a certified crane operator, but he was unfamiliar with the particular crane being used and did not know about the fifth outrigger. An operator should be qualified on the specific piece of equipment he or she will operate, as well as qualified on an equipment class (such as cranes). Equipment placards can be helpful for identifying unusual equipment characteristics.

Crane Operator Certification

As stated previously, crane operators must be trained and qualified to operate their equipment. Completion of a training course does not ensure that a worker is qualified; on-the-job training is an important supplement to formal instruction. Training isn't complete until proficiency can be consistently demonstrated.

On February 26, 1999, OSHA and the National Commission for the Certification of Crane Operators (NCCCO) signed an Agreement officially recognizing the national crane operator certification (CCO) program. Both CCO and NCCCO are trademarked acronyms of the National Commission for the Certification of Crane Operators. Either reference is correct, although NCCCO more completely refers to the organization, while CCO is generally used as the credential. The Agreement marks the first time OSHA has recognized a private sector industry group as meeting its requirements for crane operator qualification. The Agreement provides significant incentives for employers that have their operators certified through the national CCO program, as there is currently no federal requirement for crane operators to be licensed.

Assistant Secretary of Labor for OSHA, Charles Jeffress said that from his personal experience, crane operator certification is very much needed and it's very important to make sure that the folks handling this equipment are well trained. Jeffres also said that with the development of the CCO national crane operator program, OSHA is now able to say to contractors that if their crane operators have certification from the NCCCO, then OSHA will accept that as evidence of meeting the required training.

OSHA has a goal to reduce fatalities in the construction industry by 15 percent by 2002, and this agreement is seen by many as an important building block toward that end. One of the strengths of the CCO certification program is that it will separate those who have the necessary knowledge and skills from those who do not. When employees are required to be certified for a certain job, not everybody is going to pass the test. However, having a requirement for crane operator certification will cause people to get the training that is necessary to make job sites safer and help reduce the number of accidents that occur with cranes.

The signing ceremony was held at the Department of Labor in Washington, D.C., and was described as an impressive show of support by senior representatives from many sectors of industry. Labor and management, union as well as open-shop firms and organizations, and other government departments were equally represented.

Those in favor of the certification program said this signing was a great day for crane safety. This was partly because OSHA's recognition of the NCCCO crane operator certification program should have a significant and sustained impact on the safety of all who work with or around cranes. The signing of the agreement was also attributed to the dedication of hundreds of volunteers from all sectors of industry who have worked long and hard over the last 12 years to turn the dream of a national crane operator certification program into reality. Others in favor of certification also said this will go down as one of the best things ever done in the field of safety because the net result can only be a reduction in crane deaths, injuries, and accidents on construction sites around the country.

A key issue in OSHA's decision to recognize the CCO program was that it is an objective and independent assessment of the skills and knowledge of crane operators. One requirement of CCO's accreditation by the National Organization for Competency Assurance (NOCA) was that it would do no training. OSHA will continue to use crane inspection guidelines, but OSHA compliance safety and health officers, when performing inspections or accident investigations, will recognize CCO certification as verification of crane operator qualifications.

By requiring CCO-certified crane operators on a project, contractors will show a commitment to an effective safety and health program. A further benefit to employers is that the presence of

CCO-certified crane operators on a job site will indicate that cranes are being operated by employees with demonstrated knowledge and ability.

The NCCCO is an independent, not-for-profit corporation formed in January 1995 to develop effective performance standards for safe crane operation to assist all segments of general industry and construction. The program is accredited by the National Commission for Certifying Agencies (NCCA), the premier authority on certification standards. Among NCCA's exacting requirements for accreditation is that the certifying organization is third-party, independent of training; has been established and supported by industry; is a joint labor/management initiative; and administers tests that are psychometrically sound, validated through peer review, administered on a standardized basis, and maintained under strict security.

The agreement between OSHA and the NCCCO is a voluntary cooperative action between representatives of the crane industry and OSHA to recognize crane operator certification issued by NCCCO. The ability of crane operators to safely operate mobile cranes plays a major role in overall safety on most construction sites. The agreement provides a non-regulatory means of recognizing the CCO program as validating the competency and certifying the qualifications of crane operators.

When performing inspections or accident investigations, OSHA compliance safety and health officers will recognize CCO certification as verification of crane operator qualifications. The presence of CCO-certified crane operators on a job site will be indicators to compliance officers that the operator has demonstrated knowledge and ability. The presence of CCO certified crane operators on a project is an indication of the contractors' commitment to an effective safety and health program, and contributes to the project's qualification for a "focused inspection."

The CCO program meets the qualifications criteria outlined in ANSI/ASME B30.5-3.1-1995, Qualifications for and Conduct of Operators and Operating Practices. It is an objective and independent assessment of the skills and knowledge of crane operators. An applicant's certification is based on his/her performance on both the written and practical tests. The program was developed through the joint efforts of the lifting industry and labor to promote the common goal of safe crane operations.

As the CCO certification program becomes more widely used, OSHA believes education and training will become primary factors in developing and maintaining qualified crane operators in the construction industry. Certification will become the natural progression in crane operators' careers as they gain more education, training and experience working with the multitude of equipment in use today and the increasingly more advanced cranes of the future. OSHA has stated that CCO certification is based on criteria that will contribute to the mutual goal of reducing the number of deaths and injuries resulting from crane related accidents. The execution of this agreement with CCO should have an immediate, significant, and beneficial impact on safe crane operations.

THE SPECIALIZED CARRIERS AND RIGGING ASSOCIATION

Since 1959, the Specialized Carriers and Rigging Association (SC&RA), which represents most of the construction crane owners and users, has helped its members in 34 countries attain the highest standard for quality and safety through meetings, publications, services, and staff.

SC&RA members engage in a variety of activities including:

- Crane and equipment rental
- Heavy lifting
- Machinery moving and erecting
- Millwrighting
- Oil field equipment hauling
- Oversize/overweight transportation
- Plant maintenance
- Rigging
- Steel hauling
- Military material movement.

Operator Certification [9]

The national program of crane operator certification was developed and is administered by the National Commission for the Certification of Crane Operators (CCO). By providing a thorough, independent assessment of operator knowledge and skills, CCO aims to enhance lifting equipment safety, reduce workplace risk, improve performance records, stimulate training, and give due recognition to the professional skill of crane operation.

CCO was formed in January 1995 as a not-for-profit corporation to develop effective performance standards for safe crane operation to assist all segments of general industry and construction. The establishment of CCO was the culmination of almost 10 years of continuous work by a group of individuals representing all industries that use cranes. In essence, the CCO program was developed by industry for industry, and continues to be supported by industry. This industry diversity continues to be reflected through CCO's board of directors as well as its panel of commissioners, who represent such groups as contractors, labor unions, rental firms, owners, steel erectors, manufacturers, construction firms, training consultants, and insurance companies.

The wealth of crane knowledge these experts have brought to this effort has been coupled with the psychometric expertise of one of the nation's most prestigious organizations, Professional Examination Services (PES). PES played a crucial role in the development of the CCO program and continues to assist in the administration and further development of written and practical examinations.

This pooling of crane-related experience and knowledge has been supplemented with input from OSHA as well as the ANSI/ASME committees that administer the B30 crane standard. The result is a sound, valid, and effective test of the operator's proficiency. Clearly, employers, operators, and the general public have much to gain from ensuring that only qualified personnel operate cranes.

Certification

Certification is the final link in the process that educates workers in the correct way to operate cranes. Informed operators make fewer mistakes and therefore have fewer accidents than those with less or inferior knowledge.

> *Note: The CCO has been selected by NIOSH for a study of the effectiveness of crane operator certification. A two-year study will examine incidents by type of crane and type of error for each incident (e.g., operator error, power line contact, load handling, etc.). The study will test the hypothesis that there should be a decrease in the number of skills-related incidents after certification is put in place. (See Appendix K for the full text of the NIOSH press release.)*

However, certification can mean different things to different people. While certification generally involves some form of testing, not all testing qualifies as certification. This confusion can lead an employer into a false sense of security, one that could have devastating consequences in the event of an accident.

Fortunately, a set of guidelines for certification has been established by an independent credentialing authority, the National Commission for Certifying Agencies (NCCA). The NCCA is an independent not-for-profit organization set up by the National Organization for Competency Assurance (NOCA) to establish industry guidelines for professional certifying organizations.

The NCCA requirements, though strict, are designed to give assurance to those who use a program that the tests are a fair, sound, and valid measurement of the knowledge and skills they are intended to measure. Some of the requirements include:

- The certifying organization must be separate from the education function (i.e., it shall do no training).

- The certification program must be operated by a not-for-profit organization.

- The certifying organization must have a governing body that includes individuals from the discipline being certified.

So, while the CCO does not offer training, it does provide an independent means to verify that training has been effective—that learning has, in fact, taken place. Only third-party independent certification can do this, and then only if it has been validated by the industry it is intended for, and recognized as psychometrically sound by certification specialists. CCO has met all of these criteria. The key elements of the CCO program are that it:

- Actively encourages training, yet is separate from it

- Verifies that training has been effective

- Was modeled in a non-regulatory environment

- Is modeled on ANSI/ASME consensus lines

- Meets recognized professional credentialing criteria

- Has participation from all industry sectors.

Moreover, the CCO program is:

- Nationwide in scope

- Independent of labor relations policies (a labor/management joint initiative)

- Psychometrically sound (and therefore legally defensible) through CCO's partnership with PES (see next section)

- Validated through industry and peer review at all stages of development

- Administered in a standardized format, under complete security, at all times.

PROFESSIONAL CREDENTIALING

In order to ensure the CCO examination remains a psychometrically sound, valid, and effective measurement of a crane operator's knowledge and skills, CCO has teamed the expertise of its subject matter experts with a professional credentialing organization, Professional Examination Service (PES). PES played a vital role in the development of the CCO test, and continues to assist CCO in the administration and further development of written and practical examinations used in the national certification program.

This combination of crane knowledge and psychometric expertise, supplemented with input from OSHA and ANSI/ASME committees, is unprecedented in this industry.

PES is a nonprofit corporation specializing in providing licensing and certification exams. Its mission is to promote the public good by providing services in assessment, education, science, and credentialing policy.

The entire PES examination development methodology complies with the Standards for Educational and Psychological Testing and is consistent with all equal employment opportunity commission guidelines. Its examination construction process focuses on procedures, which help ensure content validity, enhancing the quality and defensibility of clients' examinations.

PES has developed content validity rating scales and rigorous item review procedures. These scales ensure that each item is tied to test specifications, at the appropriate educational level, critical to professional practice, and services a public protection function. The item review scales, tailored for each program, help content experts in screening items thoroughly. PES has found content validity to be a stronger form of legal defense than statistical information alone.

The net result of these procedures has been PES' unblemished record of legal defensibility; in the 50 years that PES has worked with its credentialing client, there has never been a successful legal challenge to the validity of any of its examinations.

About OSHA

Since 1970, OSHA has had one goal: send every worker home whole and healthy every day. Workplace fatalities have been cut in half, but nearly 17 workers still lose their lives every day, while 50 more are injured every minute during a 40-hour workweek.

Despite OSHA's accomplishments, there were still more than 6,000 fatal work injuries in 1993. In addition, every year there are still an estimated 50,000 deaths from illnesses caused by workplace chemical exposures and six million non-fatal workplace injuries. Injuries alone cost the economy more than $110 billion a year. These numbers are unacceptably high, particularly because illnesses, injuries, and deaths on the job are largely preventable.

OSHA—with a staff of 2,221, including 1,238 inspectors, and a budget of $353 million—covers more than 100 million Americans at more than six million worksites. Sharing that responsibility are 25 states that run their own OSHA programs with 2,786 employees, including 1,246 inspectors.

As our nation enters the 21st century, OSHA will look to strong enforcement, strategic management, creative partnerships, improved rulemaking, and expanded outreach and training to foster continued progress in protecting Americans on the job. The agency's top priority is developing a standard to require safety and health programs in our nation's workplaces—a proven strategy for protecting workers.

Strong Enforcement

Enforcement will continue to receive high priority, with OSHA targeting its inspections based on site-specific injury and illness data. Employers with high rates, who most need OSHA's help, are the most likely to be inspected. Top priorities for legislative changes include stronger whistleblower protections, stiffer criminal penalties when workplace deaths result from violations of OSHA standards, and coverage for all federal and state workers.

Strategic Management

OSHA recently identified 18 priority safety and health hazards in need of either regulatory or non-regulatory action. These new priorities supplement rather than replace OSHA's ongoing activities. They involve preventable problems that are responsible for a large number of injuries and illnesses (or, in some cases, that pose particularly high risks to a small group) but currently receive inadequate attention from government agencies or organizations in the private sector. Five of the new priorities have been chosen for rulemaking, while OSHA will work with business, labor, the professional community, and its state plan partners on all of the other priorities (one of which is crane hoist safety) to encourage worker protection without developing new rules at this time. In some cases, interventions may involve OSHA's use of its existing authority, as well as program initiatives recently announced by President Clinton that provide incentives to employers who effectively find and fix hazards. In most cases, the current approaches to the new priorities will be voluntary and informational.

Over the next four years, OSHA is committed to reducing workplace injuries and illnesses in three target groups:

- In the 100,000 workplaces where they conduct major interventions (20% reduction)
- In five high-hazard industries—food processing, nursing homes, shipyards, logging, and construction (15% reduction)
- In three serious safety and health problems—silicosis, amputations, and lead poisoning (15% reduction).

Creative Partnerships

Companies that demonstrate excellence in workplace safety and health may request OSHA recognition under the agency's premier partnership—the Voluntary Protection Program (VPP). In 1999, nearly 500 workplaces participated. Together, VPP sites in more than 180 industries are saving nearly $135 million each year because their injury rates are almost 60% below the average for their industry.

Specialized partnerships focusing on residential construction, steel erection, roofing, and other issues have also made strides in reducing workplace injuries and illnesses. The common thread through all these partnerships is the commitment of each organization and participant to work cooperatively with OSHA to implement effective safety and health programs. However, OSHA's pilot partnership projects in Maine and Wisconsin have drawn toward excellence participants that had previously had less than stellar records.

Improved Rulemaking

New OSHA rules will emphasize outcomes and results rather than specific requirements. These rules will provide ongoing protection in a constantly changing environment. OSHA standards must offer flexibility to enable employers to continue to reduce hazards and avoid injury and illness as the workplace evolves. To get rules out more quickly, OSHA is involving stakeholders early on, and is using teams to speed its internal rulemaking process.

Expanded Education and Outreach

OSHA is planning an extensive outreach program to assist employers in developing safety and health programs and addressing ergonomic hazards. The agency is also working to improve training for inspectors to help them learn to evaluate workplace safety and health programs. System analysis, as opposed to identifying specific hazards during inspections, represents a new approach—a culture change for the agency. In addition, OSHA is seeking to codify its free consultation program that helps employers find and fix hazards and establish safety and health programs.

OSHA Inspections

At the beginning of fiscal year 1999, OSHA had 2,221 full-time employees and a budget appropriation of $353 million. During the fiscal year 1998, Federal OSHA conducted 34,442 inspections, while state inspections for the same period totaled 55,699.

The reason for inspections were:	Inspections by industry sector were:
Federal inspections:	Federal inspections:
• Programmed: 47 percent	• Construction industry: 53 percent
• Complaint: 23 percent	• Manufacturing industry: 26 percent
• Referrals/follow-ups: 27 percent	• Other industries: 21 percent
• Accident: 3 percent	State inspections:
State inspections:	• Construction industry: 43 percent
• Programmed: 56 percent	• Manufacturing industry: 22 percent
• Complaint: 21 percent	• Other industries: 35 percent
• Referrals/follow-ups: 16 percent	
• Accident: 7 percent	

In the inspections categorized above, Federal OSHA identified 76,980 violations, with total penalties of $79,996,761. For the same categories, State OSHA identified 149,513 violations, with total penalties of $53,086,045.

OSHA's Goal

OSHA's primary goal is to ensure safe and healthful working conditions for every American worker. All of the agency's enforcement, educational, and partnership efforts seek to reduce the number of occupational injuries, illnesses, and deaths. The emphasis is on results—sending every worker home whole and healthy every day.

Worker Injuries/Illness 1996—6.2 million injuries/illnesses among private sector firms with 11 or more employees was about 400,000 less than in 1995. The rate of injuries and illnesses for every 100 workers dropped from 8.9 in 1992 to 7.4 in 1996—the lowest on record. The rate for serious cases—2.2—also was the lowest on record. The 6.2 million cases included 5.8 million injuries and 459,000 illnesses, of which 64% were associated with repeat trauma.

Worker Fatalities 1997—6,218—40% transportation related; 18% caused by violence. Construction industry was highest with 18% fatalities and 6% of total employees. [3]

OSHA, in most cases, cannot impose on employers more stringent safety and health requirements than promulgated in OSHA standards. However, OSHA traditionally encourages employers to meet current national consensus standards, which meet or exceed OSHA standards, to enhance the safety and health of employees in the workplace. An example is a more recent ANSI standard applicable to construction cranes which may be cited by OSHA when the applicable OSHA standard references an earlier version of the ANSI standard. This is possible even though there is no automatic adoption of the later standard by OSHA. However, employers not only must comply with specific OSHA standards but also have an obligation under the general duty clause of the Act, Section 5(a)(1). The general duty clause may be cited when a particular hazard is not addressed by any standard or when the OSHA standard is known to be insufficiently protective. An example would be that if OSHA issued a citation for lack of a boom length or angle indicator, the ANSI requirements would likely be part of the evidence, showing that a recognized hazard exists and that boom length and angle indicators are appropriate means to eliminate or reduce the hazard.

There is no specific OSHA standard that requires that a crane constructed and installed before August 31, 1971, conform to the specifications of ANSI B30.2196. However, it is possible that OSHA could issue a citation under section 5(a)(1) of the Occupational Safety and Health Act (the general duty clause) if such a crane presents a hazard that is serious and recognized within the industry or by safety experts familiar with the industry. The age of a crane might have some bearing on the issue of feasible abatement, especially since the current 1991 version of ANSI B30.2 continues to "grandfather" pre-1971 cranes.

If a particular hazard is presented that is not addressed by any specific OSHA standard, a citation under the general duty clause may be appropriate. An ANSI standard such as B30.1 might be used by OSHA to establish both recognition of the hazard and the existence of a feasible means of abatement. Although OSHA sometimes makes use of a consensus standard as evidence to support a citation in litigation, this does not mean that the consensus standard has become an OSHA standard having the force of law. In instances where OSHA standards do not specifically address a particular activity or hazard, ANSI standards can and have been used by OSHA as a basis for a citation under 5(a)(1) of the Act.

CRANE SAFETY RATIONALE

Crane and hoist safety meets several of the criteria for designation as an OSHA priority:

- The very serious nature of the hazard

- The magnitude of the risk (a high rate of fatalities and serious injuries relative to the number of workers exposed)

- The potential for catastrophic accidents.

The considerable knowledge about effective protective measures clearly demonstrates the need for action to address crane and hoist safety.

The knowledge of a supervisory employee of a violative condition is properly imputed (transmitted) to the employer. The test for knowledge would be whether an employer knew, or, with the exercise of reasonable diligence, could have known of the presence of a violative condition. An example would be that an employer couldn't say they did not have knowledge of a violative condition because the project did not call for working near electrical wires at that time. The fact that a company does not intend for its crew to operate a crane near electrical wires does not excuse a supervisor's failure to exercise caution and inspect his worksite when such situations arise. Reasonable diligence includes "the obligation to inspect the work area, to anticipate hazards to which employees may be exposed, and to take measures to prevent the occurrence."

A serious violation is defined as one in which there is a substantial probability that death or serious physical harm could result, and the employer knew or should have known of the hazard. A willful violation is one committed with intentional, knowing or voluntary disregard for the requirements of OSHA regulations, or with plain indifference to employee safety. The employer is responsible for the willful nature of its supervisors' actions to the same extent that the employer is responsible for their knowledge of the following violative condition:

> *A crew foreman knew that an electrical line was not covered as an OSHA compliance officer had said was required, yet the foreman ordered the crew to continue to work in close proximity of the line before protective equipment arrived. This would be a willful violation because the supervisor had received warnings of a serious hazard, and it was not corrected. An employer is responsible for the willful nature of a foreman's disregard of instructions where the foreman's action is preventable.*

To establish a violation, OSHA must show that:

- The standard is applicable
- The employer failed to comply with it
- Employees had access to the violative condition
- The employer had knowledge or constructive knowledge of the condition.

In order to prove a violation of section 5(a)(1) of the Act, OSHA must show that:

- A condition or activity in the workplace presented a hazard to an employee

- The hazard was recognized

- The hazard was likely to cause death or serious physical harm

- A feasible means existed to eliminate or materially reduce the hazard.

The evidence must show that the employer knew, or with the exercise of reasonable diligence could have known, of the violative conditions. Generally, an employer cannot rely on the failure of OSHA to issue a citation for a particular condition during an earlier inspection as the basis for later arguing lack of knowledge of the same hazardous condition.

To prevail on the unpreventable employee misconduct defense, an employer must show that it had "established work rules designed to prevent the violation." This would include adequately communicating those work rules to its employees (including supervisors), taking reasonable steps to discover violations of those work rules, and effectively enforcing work rules when they were violated.

EXAMPLES OF OSHA CITATIONS

In fiscal year 1998, OSHA issued 79,980 citations, with penalties totaling $79,996,761. Included here are examples of crane-related citations from prior years.

After a fatal accident at a Naval Shipyard on August 24, 1994 aboard an aircraft carrier, OSHA issued a citation to the Shipyard for violation of 29 CFR 1915.117(b). The citation stated that "employees who did not understand the signs, notices, and operating instructions were permitted to operate a crane, winch, or other power-operated hoisting apparatus."

Other crane-related citations have been issued by OSHA for:

- Failure to perform complete periodic inspections of overhead cranes and unsafe conditions not corrected before placing cranes back in service after inspections

- Operating semi-gantry crane on rails that were not adequately spliced, did not have smooth joints, and were not properly secured to foundations

- Using custom-made rigging devices not marked with rated capacities, nor proof tested

- Operating an overhead cab-operated crane not equipped with a warning signal

- Serious violations for which one employer was cited after a crane operator and another worker were injured when an overloaded crane cable broke, dropping its boom, included:

  - Failure to post a crane load capacity chart and crane hand signal chart

  - Failure to inspect the crane

  - Failure to follow the crane manufacturer's load capacity restrictions

  - Failure to provide fall protection and fall protection training

  - Failure to properly set the height of guardrails, to provide mid-rails, and to provide railings capable of supporting the required 200-pound minimum

  - Failure to have a hazard communication program, including employee training and material safety data sheets covering the hazardous substances in the workplace

  - Failure to conduct overall site-specific safety training

  - Failure to provide protective gloves

  - Failure to store liquefied petroleum gas cylinders upright, and separately from oxygen cylinders.

As shown here, rules violations during crane operations can lead to additional citations by OSHA.

SUMMARY

The potential exists for a serious accident in crane-related tasks that could result in consequential property damage, personnel injury, or indeed death. Because of the possibility of such an event occurring, it is essential that employers be compliant with all the current crane-related rules and regulations.

OSHA's analysis of crane accidents over a five-year period identified the major causes as:

- Boom or crane contact with energized power lines (nearly 45% of the cases)

- Under-the-hook lifting device failure

- Overturned cranes

- Dropped loads

- Boom collapse

- Crushing by the counterweight

- Improper outrigger use

- Falls

- Rigging failures.

All of these accidents could have been avoided with properly trained operators/riggers and the use of equipment that was reliable, and properly designed, inspected, and maintained.

In an attempt to focus employers' commitment to an effective safety and health program, OSHA signed an agreement with the NCCCO officially recognizing the national crane operator certification program. The agreement marks the first time OSHA has recognized a private sector industry group as meeting its requirements for crane operator qualifications. The agreement provides significant incentives for employers that have their operators certified through the national CCO program as there is currently no federal requirement for crane operators to be licensed.

By fully implementing the appropriate OSHA requirements and ANSI/ASME standards, employers can significantly reduce the likelihood of events causing personal injury or property damage. To reduce and eliminate such events, the safety of employees must take precedence on all jobs.

REFERENCES

1. OSHA, (September 1990). Draft Report by OSHA's Crane Safety Task Group.

2. BLS (1995). *Bureau of Labor Statistics 1993 Census of Fatal Occupational Injuries.*

3. (57 CFR 47746) Federal Register, October 19, 1992. Occupational Safety and Health Administration, Crane Safety for General Industry and Construction, Advance Notice of Proposed Rulemaking.

4. Pratt, S. G., Kisner, S. M., Moore P. H., The NTOF surveillance system. *Machinery-related Fatalities in the U.S. Construction Industry*, 1980-1992.

5. NIOSH (1995). NIOSH Alert: Preventing Electrocutions of Crane Operators and Crew Members Working Near Overhead Power Lines. (NIOSH Publication No. 95-108). May 1995.

6. ANSI (1994). American National Standards Institute B30 Series Standards.

7. Crane Safety, *Occupational Safety Observer*, EH-9308 Issue No. 8, August 1993, pp. 1-2.

8. "OSHA Signs Agreement With Crane Operator Certification Program," U.S. Department of Labor, Washington, D.C., February 26, 1999.
   Available from http://www.nccco.org/whatsnew/osha.htm

9. Specialized Carriers and Rigging Association (SC&RA). "Welcome to Operator Certification", Available from http://www.scranet.org/ccorod.html.

10. Virginia Considers Crane Operator Certification, *CraneWorks,* March-April 1998, page 11.

11. U.S. Department of Labor, "About OSHA", Washington, D.C., February 26, 1999.
    Available from www.osha-slc.govoshafacts/osha99.pdf.

12. U.S. Department of Labor, "OSHA Facts", Washington, D.C., February 26, 1999.
    Available from www.osha-slc.govoshafacts/osha99.pdf.

2

GENERAL REQUIREMENTS

The need for action by employers to address the safe operation of cranes is established by the following issues:

- The serious nature of the hazards involved

- The magnitude of the risk (high rate of fatalities and serious injuries relative to the number of workers exposed)

- The potential for catastrophic consequences due to accidents

- The considerable knowledge available about effective protective measures.

Responsibility for the many facets of a crane or derrick operation needs to be clearly defined to everyone involved before any work begins. This will ensure continuity in the work process if an incident requiring investigation occurs. The need for identifying responsibilities extends from the employer, to the crane operator, to the employee(s) rigging the load to be lifted. Table 2.1 provides questions to help determine responsibilities in planning a lift. Table 2.1 also provides basic sling operating practices as dictated by ANSI B30.9.

Table 2.1. Responsibilities [2]

THE BASIC RIGGING PLAN

1. WHO IS RESPONSIBLE (COMPETENT) FOR THE RIGGING?
   COMMUNICATIONS ESTABLISHED?
2. IS THE EQUIPMENT IN ACCEPTABLE CONDITION?
   APPROPRIATE TYPE,
   PROPER IDENTIFICATION?
3. ARE THE WORKING LOAD LIMITS ADEQUATE?
   WHAT IS WEIGHT OF LOAD?
   WHERE IS THE CENTER OF GRAVITY?
   WHAT IS THE SLING ANGLE?

   WILL THERE BE ANY SIDE LOADING?
   CAPACITY OF THE GEAR?
4. WILL THE LOAD BE UNDER CONTROL?
   TAG LINE AVAILABLE?
   IS THERE ANY POSSIBILITY OF FOULING?
   CLEAR OF PERSONNEL?
5. ARE THERE ANY UNUSUAL LOADING OR ENVIRONMENTAL CONDITIONS?
   WIND, TEMPERATURE, OTHER?
6. YOUR SPECIAL REQUIREMENTS?

THE USERS RESPONSIBILITIES

UTILIZE APPROPRIATE RIGGING GEAR SUITABLE FOR OVERHEAD LIFTING
UTILIZE THE RIGGING GEAR WITHIN INDUSTRY STANDARDS AND THE MANUFACTURER'S
   RECOMMENDATIONS
CONDUCT REGULAR INSPECTION AND MAINTENANCE OF THE RIGGING GEAR

BASIC SLING OPERATING PRACTICES ANSI B30.9

WHENEVER ANY SLING IS USED, THE FOLLOWING PRACTICES SHALL BE OBSERVED.
1. SLINGS THAT ARE DAMAGED OR DEFECTIVE SHALL NOT BE USED.
2. SLINGS SHALL NOT BE SHORTENED WITH KNOTS OR BOLTS OR OTHER MAKESHIFT DEVICES.
3. SLING LEGS SHALL NOT BE KINKED.
4. SLINGS SHALL NOT BE LOADED IN EXCESS OF THEIR RATED CAPACITIES.
5. SLINGS USED IN A BASKET HITCH SHALL HAVE THE LOADS BALANCED TO PREVENT SLIPPAGE.
6. SLINGS SHALL BE SECURELY ATTACHED TO THEIR LOAD.
7. SLINGS SHALL BE PADDED OR PROTECTED FROM THE SHARP EDGES OF THEIR LOADS.
8. SUSPENDED LOADS SHALL BE KEPT CLEAR OF ALL OBSTRUCTION.
9. ALL EMPLOYEES SHALL BE KEPT CLEAR OF LOADS ABOUT TO BE LIFTED AND OF SUSPENDED LOADS.
10. HANDS OR FINGERS SHALL NOT BE PLACED BETWEEN THE SLING AND ITS LOAD WHILE THE SLING IS BEING TIGHTENED AROUND THE LOAD.
11. SHOCK LOADING IS PROHIBITED.
12. A SLING SHALL NOT BE PULLED FROM UNDER A LOAD WHEN THE LOAD IS RESTING ON THE SLING.

INSPECTION: EACH DAY BEFORE BEING USED, THE SLING AND ALL FASTENINGS AND ATTACHMENTS SHALL BE INSPECTED FOR DAMAGE OR DEFECTS BY A COMPETENT PERSON DESIGNATED BY THE EMPLOYER. ADDITIONAL INSPECTIONS SHALL BE PERFORMED DURING SLING USE WHERE SERVICE CONDITIONS WARRANT. DAMAGED OR DEFECTIVE SLINGS SHALL BE IMMEDIATELY REMOVED FROM SERVICE.

EMPLOYER RESPONSIBILITIES

Because the knowledge of a supervisory employee is imputed to the employer, employers are responsible for ensuring that all hoisting equipment is operated in a safe manner and only operated by qualified personnel. An example of one situation would be that before the hoisting of personnel on personnel platforms on load lines of cranes or derricks is permitted, employers must preplan the work to be done, including proof testing the rigging to be used and conducting a trial lift.

OSHA 29 CFR 1926.550(a)(5) requires employers to designate a competent person to inspect machinery and equipment prior to each use, and during use, to make sure it is in safe operating condition. Any deficiencies shall be repaired, or defective parts replaced, before continued use. 1926.550(a)(6) requires a thorough, annual inspection of the hoisting machinery to be made by a competent person, or by a government or private agency recognized by the U.S. Department of Labor. In addition, the employer is required to maintain a record of the dates and results of inspection for each hoisting machine and piece of equipment.

A competent person is described in OSHA 29 CFR 1926.32(f) as one who is capable of identifying existing and predictable hazards in the surroundings or working conditions which are unsanitary, hazardous, or dangerous to employees, and who has authorization to take prompt corrective measures to eliminate them. An example would be if employees meet the qualifications for a "competent person," they can perform the construction inspections required in 1926.550. This is because OSHA construction standards do not require employees performing crane inspections to have a rating, such as a Level II rating, which is a term used in an ANSI standard and not referenced in the OSHA standards.

Employees who operate cranes or perform other work governed by the OSHA standards shall be trained and qualified to the level of proficiency consistent with assigned duties. Employers are responsible for work assignments and must ensure that such assignments do not exceed personnel qualifications. The following set of responsibilities and training requirements can be used as a guide in most circumstances when cranes are used in the performance of work.

An employer is responsible for the operation, inspection, maintenance, and repair of equipment and components and is required to ensure that:

- A crane inspection and maintenance program is established
- Equipment is inspected, maintained, and repaired only by qualified personnel and all service is documented

- All inspection, maintenance, and repair activities are conducted and documented in accordance with the requirements of any manual furnished by the manufacturer

- Cranes are operated only by qualified personnel (i.e., qualified operators, supervised trainees, and qualified maintenance, inspection, and test personnel)

- The crane and any associated equipment are in concurrence with all regulations and manufacturer's requirements

- Any equipment found to be unsafe or requiring restrictive use is properly tagged and taken out of service

- The employee assigned to operate the crane is qualified through appropriate experience and training and is capable of carrying out those responsibilities

- Appropriate qualified supervision is provided for all jobs with authority necessary to exercise assigned responsibilities

- All employees assigned to the job understand their individual responsibilities and authority and are competent to handle their jobs safely and efficiently

- All accidents and events receive prompt attention and are investigated thoroughly to prevent future recurrences.

OPERATOR RESPONSIBILITIES

Operator responsibilities and required qualifications are driven by OSHA statistics showing that more than 90% of crane and rigging accidents are caused by human error and not equipment failure. Using equipment outside its designed criteria and lifting unstable loads are the main cause of accidents. The correct use, operation, and maintenance of cranes and related equipment will enhance the safety of those employees who work with and around such equipment.

OSHA 1910.179(b)(8) provides guidance on which employees are allowed to operate a crane. "Designated personnel—Only designated personnel shall be permitted to operate a crane covered by this section." While 1910.179 does not include any physical qualifications for overhead and gantry crane operators, an employer has the responsibility to determine whether crane operators can safely perform their work. An employer's decision could be influenced by working conditions such as operators operating a derrick strictly by headset communications or crane operators having visual problems in following the signals provided by a signalman. The physical qualifications required in the ANSI B30.17-1980 standard are advisory requirements, which have not been adopted by OSHA and cannot be enforced on operators of equipment covered by 29 CFR 1910.179.

The use of the general duty clause 5(a)(1) of the OSHA act is warranted only when there is a substantial probability that a recognized hazard may cause death or serious physical harm to employees. One example is whether the use of a crane operator who has limited vision in one eye presents such a recognized hazard, and the answer would depend on the particular work situation in which the person is operating.

Whenever a crane or derrick is used in its regular production activities, such as material handling and loading, and is under the control of its operator rather than under the control of a maintenance worker, it is considered to be used in a normal operating condition. However, *normal operating conditions* do not encompass the situation when maintenance is performed on a crane or derrick which is taken out of production and the crane or derrick is under the control of the maintenance worker.

The term "normal operating conditions" is used in 29 CFR 1910.179(e)(6)(i), 1910.179(g)(2)(i), 1910.181(j)(1)(i), 29 CFR 1926.550(d)(4) (ANSI B30.2.0.-1967, Chapter 2-1.7.7, Paragraph A and Chapter 2-1.9.2, Paragraph A) and 29 CFR 1926.550(e) (ANSI B30.6, 1960, Chapter 6-1.6.1, Paragraph A).

The crane operator is responsible for those operations under the operator's direct control. If there is ever any doubt as to safety, it is the responsibility of the operator to consult with the foreman or supervisor before continuing rigging or lifting the loads. Some operator responsibilities are:

- Ensuring no one enters a crane cab or pulpit when the crane is in operation

- Safely assembling, setting up, and operating all assigned equipment properly

- Following the equipment operating guidelines specified in the crane's operating manual and, for mobile cranes, the load charts

- Knowing the weight of the load and associated rigging

- Performing the pre-use and frequent equipment inspection

- Fully understanding the crane's capacity, functions, and operational limitations and ensuring that the load will not exceed the rated capacity of the equipment

- Responding to signals from a designated person directing the lift, except for the stop signal, which must be obeyed any time anyone gives it (If a signal person is not required as part of the lifting operation, the operator assumes total responsibility for the lift.)

- Abiding by any restrictions placed on the use of the equipment

- Making sure there is no load attached before leaving the crane unattended

- Securing the crane when leaving it unattended.

RIGGER RESPONSIBILITIES / GUIDELINES

Only qualified and authorized employees should be permitted to perform rigging functions. A qualified rigger is defined in the Glossary as "one whose competence in this skill has been demonstrated by experience satisfactory to the employer or appointed person." The following activities are identified as some of the prescribed responsibilities of those employees performing such rigging assignments.

It is the responsibility of the rigger to:

- Ensure that the rigging equipment and materials have the required capacity for the job and that all items are in good condition, are currently qualified (inspection is up-to-date), and are properly used

- Verify that rigging equipment and material are in compliance with any applicable procedure

- Confirm that the load path is clear of personnel and obstacles

- Have knowledge of the operating characteristics of the equipment in use, know its capacity, limitations, and condition

- Verify the stability of the equipment.

It is also the responsibility of the assigned rigger(s) to be aware of or have knowledge of:

- Potential environmental hazards such as the weather

- The potential dangers of electrical hazards

- Load weight estimation

- Emergency procedures

- Rigging equipment selection

- Hand signals

- Maintenance/storage of slings and rigging components

- Load dynamics

- Applicable standards and regulations

- Safety features of equipment

- Terminology and definitions

- Ropes and reeving

- Rigging/operating practices

- Sling loading

- Load-indicating devices

- Personal protective equipment

- Below-the-hook lifting devices

- Rigging or hitch configuration

- D/d ratio

- Sling types and application.

RIGGER GUIDELINES

Before selecting a sling for a specific list, it must be determined which hitch is the most effective to do the job, protect the load, and protect the sling. One of three basic hitches will usually do the job. The type of hitch selected may determine the type of sling body that will best do the job, as well as the length of the sling that will be needed. Lifting height, overhead clearance, and hook travel will affect the choice of hitch and length of the sling.

A sling body type should be chosen that will best support the load while providing adequate rated capacity. The proper choice will provide:

- Lifting capacity needed

- Proper D/d ratio

- Handling characteristics needed for rigging

- Minimal damage to the sling

- Minimal damage to the load.

Additional useful guidelines for riggers include:

- Use a spreader bar between legs of a sling to prevent pressure on the load by the sling during the lift.

- When attaching a sling to eyebolts, always pull on the line with the bolt axis. When hitching to bolts screwed into or attached to a load, a side pull may break the bolts.

- Use a shackle in the eye during a choke to protect sling body against excessive distortion. Always put a shackle pin through the pin eye, rather than against the sling body since sliding movement of the sling body could rotate the pin, causing it to come loose. See Figure 3.5 on page 69 for rated capacity adjustment required when choke is made at various angles.

- A sliding hook choker is superior to a shackle or unprotected eye, since it provides a greater bending radius for the sling body.

- Use blocking or padding to protect hollow vessels, loose bundles, and fragile items from scuffing and bending. Remember that blocking becomes part of the lift and must be added to the total weight of the sling.

- When lifting crates or wooden boxes with a basket hitch, be sure the load can withstand side pressure as tension is applied to sling. Use spreader bars and corner protectors to prevent damage to load contents.

- When lifting a bundled load with a single sling near the center of gravity, a choke is more effective than a basket hitch to prevent unbalance and slipping off the load in a sling.

- You can reduce the angle of a choke with a wooden block, or blocks, between the hitch and the load. This also increases the angle between the two legs to improve sling capacity.

- When rigging two or more straight slings as a bridle, select identical sling constructions of identical length—with identical previous loading experience. Normal stretch must be the same for paired slings to avoid overloading individual legs and unbalancing the load during the lift.

OSHA 1910 SUBPART N

The following applies to the three sections (overhead and gantry cranes, mobile cranes, and derricks) in OSHA 1910 Subpart N. Included are the requirements for the operation, inspection, maintenance, and testing requirements for each.

Overhead and gantry cranes include semigantry, cantilever gantry, wall cranes, storage bridge cranes, and others having the same fundamental characteristics. These cranes may be top-running, under-running, or single- or double-girder. Overhead, gantry, and mobile cranes may be cab operated, pulpit operated, floor operated, or remotely operated. Such cranes are grouped together because all have trolleys and similar travel characteristics and are governed by OSHA regulation, 29 CFR 1910.179.

Mobile cranes pertain to crawler cranes, locomotive cranes, wheel-mounted cranes, and any variations that retain the same fundamental characteristics. These cranes rotate 360 degrees and have boom-luffing capabilities and are governed by OSHA regulation 29 CFR 1910.180.

Section 1910.181 applies to guy, stiffleg, basket, breast, gin pole, Chicago boom, and A-frame derricks of the stationary type. They are capable of handling loads at various reaches and are powered by hoists through systems of rope reeving, used to perform lifting work, single, or multiple line bucket work, grab, grapple, and magnet work. Derricks may be permanently installed for temporary use in construction work.

It should be noted that OSHA construction regulations provide minimum safety and health standards for employees working in the construction industry. However, more stringent requirements, such as an ANSI/ASME standard for crane inspections at nuclear power plants, may be required by virtue of the contractual relationship for construction at the facility. A list of certification agencies accredited by the Occupational Safety and Health Administration to perform crane inspections is available from OSHA.

Inspections (Overhead/Gantry Cranes, Mobile Cranes, and Derricks)

There are currently no OSHA regulations requiring the certification of cranes, derricks, and material handling devices used solely in general industry operations (covered under 29 CFR Part 1910), or used solely in construction operations (covered under 29 CFR Part 1926). Therefore, cranes, derricks, and material handling devices used exclusively in general industry or construction operations are not required under OSHA regulations to be certificated by anyone. The owner must, however, maintain a record of inspections. OSHA regulations only require that such equipment be inspected during initial use, frequently, and annually thereafter by a "competent person." A competent person does not have to be a disinterested third party. A competent person can be the equipment owner's maintenance personnel or any other person the owner chooses, as long as that person is deemed "competent."

Prior to initial use and at other times as required, all new and altered cranes shall be inspected to ensure compliance with the following OSHA and ASME standards.

Overhead Cranes

- ASME B30.2, "Overhead and Gantry Cranes"
- OSHA 1910.179, "Overhead and Gantry Cranes" (see Chapter 4)

Mobile Cranes

- ASME B30.5, "Mobile and Locomotive Cranes"
- OSHA 1910.180, "Crawler, Locomotive, and Truck Cranes" (Page 126)
- OSHA 1926.550, "Cranes and Derricks." (Page 162)

Derricks

- ASME B30.6, "Derricks"
- OSHA 1910.181, "Derricks" (Page 146)

OVERHEAD / GANTRY CRANE INSPECTION

Initial Inspection

Before initial use, new, reinstalled, altered, modified, or extensively repaired overhead and gantry cranes must be inspected in accordance with a written procedure. This inspection should include the following functions:

- Hoisting and lowering
- Trolley travel
- Bridge travel
- Limit switches and locking and safety devices.

A record of the initial inspection shall be prepared by a qualified inspector and retained in the crane history file.

In addition to the initial inspections, frequent inspections are required on daily to monthly intervals and periodic inspection on one to 12-month intervals depending on the crane's activity, severity of service, and environmental conditions. The inspection shall include the following areas:

- Deformed, cracked, corroded, worn, or loose members or parts
- The brake system
- Limit indicators (wind, load)
- Power plant, and electrical apparatus.

The reports must be dated and signed as with the initial inspection reports and retained in the crane history file.

OSHA 1910.179(j)(2) provides the following guidance on daily inspections of overhead cranes:

Frequent inspection. The following items shall be inspected for defects at intervals as defined in paragraph (j)(1)(ii) of this section or as specifically indicated, including observation during operation for any defects that might appear between regular inspections. All deficiencies such as listed shall be carefully examined and determination made as to whether they constitute a safety hazard:

- (j)(2)(i) All functional operating mechanisms for maladjustment interfering with proper operation. Daily.
- (j)(2)(ii) Deterioration or leakage in lines, tanks, valves, drain pumps, and other parts of air or hydraulic systems. Daily.

Frequent Inspection – Daily to Monthly (Checklist #1 in Appendix M)

It is the duty of the operator to inspect the crane each day of use for deficiencies that may have occurred between regular inspections. This would include visual inspections of the functional operating mechanisms, air and hydraulic systems, chains, ropes, slings, hooks, and other lifting equipment. All motions (braking, hoisting, etc.) must be smooth and regular with no hesitations, vibration, binding, weaving, unusual noise, or other irregularity. Any deterioration or leakage in lines, tanks, valves, drain pumps, and other parts of air or hydraulic systems must be corrected before the crane is operated.

Periodic Crane Inspection Requirements (Checklist #2 in Appendix M)

For periodic crane inspection requirements, OSHA standard 1910.179(j)(3), specifically states:

Complete inspections of the crane shall be performed at intervals as generally defined in paragraph (j)(1)(ii)(b) of this section, depending upon its activity, severity of service, and environment, or as specifically indicated below. These inspections shall include the requirements of paragraph (j)(2) of this section and, in addition, the following items. Any deficiencies such as listed shall be carefully examined and determination made as to whether they constitute a safety hazard:

- (i) Deformed, cracked, or corroded members
- (ii) Loose bolts or rivets
- (iii) Cracked or worn sheaves and drums
- (iv) Worn, cracked, or distorted parts such as pins, bearings, shafts, gears, rollers, locking and clamping devices
- (v) Excessive wear on brake system parts, linings, pawls, and ratchets {1910.179(j)(3)(vi)}

- (vi) Load, wind, and other indicators over their full range, for any significant inaccuracies

- (vii) Gasoline, diesel, electric, or other power plants for improper performance or non-compliance with applicable safety requirements

- (viii) Excessive wear of chain drive sprockets and excessive chain stretch.

Mobile Crane Inspection

Before initial use, new and altered cranes are required to be inspected by a qualified inspector to ensure compliance with the following ASME and OSHA regulations:

- ASME B30.5, "Mobile and Locomotive Cranes"

- OSHA 1910.180, "Crawler Locomotive and Truck Cranes"

- OSHA 1926.550, "Cranes and Derricks"

Dated and signed initial inspection reports shall be retained in the crane history file.

For frequent mobile crane inspections *(Checklist #3 in Appendix M)* OSHA Standard 1910.180(d)(3) specifies the following:

"Frequent inspection." Items such as the following shall be inspected for defects at intervals as defined in subdivision (2)(i) of this subparagraph or as specifically indicated, including observation during operation for any defects that might appear between regular inspections. Any deficiencies listed shall be carefully examined and determination made as to whether they constitute a safety hazard:

- (i) All control mechanisms for maladjustment interfering with proper operation: Daily.

- (ii) All control mechanisms for excessive wear of components and contamination by lubricants or other foreign matter.

- (iii) All safety devices for malfunction.

- (iv) Deterioration or leakage in air or hydraulic systems: Daily.

- (v) Crane hooks with deformations or cracks. For hooks with cracks or having more than 15% in excess of normal throat opening or more than 10 degree twist from the plane of the unbent hook.

- (vi) Rope reeving for noncompliance with manufacturer's recommendations.

- (vii) Electrical apparatus for malfunctioning, signs of excessive deterioration, dirt, and moisture accumulation.

For periodic mobile crane inspections *(Checklist #4 in Appendix M)*, OSHA Standard 1910.180(d)(4) specifies the following:

"Periodic inspection." Complete inspections of the crane shall be performed at intervals as generally defined in paragraph (d)(2)(ii) of this section depending upon its activity, severity of service, and environment, or as specifically indicated below. These inspections shall include the requirements of paragraph (d)(3) of this section and, in addition, items such as the following. Any deficiencies such as listed shall be carefully examined and determination made as to whether they constitute a safety hazard:

- (i) Deformed, cracked, or corroded members in the crane structure and boom

- (ii) Loose bolts or rivets

- (iii) Cracked or worn sheaves and drums

- (iv) Worn, cracked, or distorted parts such as pins, bearings, shafts, gears, rollers, and locking devices

- (v) Excessive wear on brake and clutch system parts, linings, pawls, and ratchets

- (vi) Load, boom angle, and other indicators over their full range, for any significant inaccuracies

- (vii) Gasoline, diesel, electric, or other powerplants for improper performance or non-compliance with safety requirements

- (viii) Excessive wear of chain-drive sprockets and excessive chain stretch

- (ix) Travel steering, braking, and locking devices for malfunction

- (x) Excessively worn or damaged tires.

Derrick Inspections

Derrick inspection requirements located in OSHA 1910.181(d)(3)(i) and (d)(3)(ii) specify the following"

Periodic inspection: 1910.181(d)(3)(i) and 1910.181(d)(3)(i).

1910.181(d)(3)(i)—Complete inspections of the derrick shall be performed at intervals as generally defined in paragraph (d)(1)(ii)(b) of this section depending upon its activity, severity of service, and environment, or as specifically indicated below. These inspections shall include the requirements of paragraph (d)(2) of this section and, in addition, items such as the following deficiencies shall be carefully examined and a determination made as to whether they constitute a safety hazard:

(d)(3)(i)(a)

Structural members for deformations, cracks, and corrosion.

(d)(3)(i)(b)

Bolts or rivets for tightness.

(d)(3)(i)(c)

Parts such as pins, bearings, shafts, gears, sheaves, drums, rollers, locking and clamping devices, for wear, cracks, and distortion.

(d)(3)(i)(d)

Gudgeon pin for cracks, wear, and distortion each time the derrick is to be erected.

(d)(3)(i)(e)

Powerplants for proper performance and compliance with applicable safety requirements.

(d)(3)(i)(f)

Hooks.

(d)(3)(ii)

Foundation or supports shall be inspected for continued ability to sustain the imposed loads.

OPERATIONAL TESTS

Before initial use, new, reinstalled, altered, repaired, or modified derricks shall be tested by a person qualified to ensure that the derrick is in good operating condition, including the following functions:

- Load lifting and lowering
- Boom up and down
- Swing
- Operation of clutches and brakes of hoist
- Limit devices, if provided.

CAPACITY AND INSPECTION OF HOOKS AND HOIST CHAINS

Hook Capacity

For the capacity of hooks, OSHA 1910.184(d)(2)(i) states:

"Hooks, rings, oblong links, pear-shaped links, welded or mechanical coupling links, or other attachments shall have a rated capacity at least equal to that of the alloy steel chain with which they are used or the sling shall not be used in excess of the rated capacity of the weakest component."

Hook Inspection

Hook inspections must be performed at different intervals based upon general classifications: before initial use, daily, frequent, and periodic.

Before initial use, all new and repaired hooks must be inspected to ensure compliance with the applicable provisions of ASME B30.10, Hooks and OSHA 1910.179(j)(2)(iii) and (1)(3)(iii)(a).

1910.179(j)(2)(iii)

Hooks with deformation or cracks. Visual inspection daily; monthly inspection with a certification record which includes the date of inspection, the signature of the person who performed the inspection and the serial number, or other identifier, of the hook inspected. For hooks with cracks, or having more than 15% in excess of normal throat opening or more than 10% twist from the plane of the unbent hook, refer to paragraph (1)(3)(iii)(a) of this section.

1910.179(l)(3)(iii)(a)

Crane hooks showing defects described in paragraph (j)(2)(iii) of this section shall be discarded. Repairs by welding or reshaping are not generally recommended. If such repairs are attempted, they shall only be done under competent supervision, and the hook shall be tested to the load requirements of paragraph (k)(2) of this section before further use.

As shown in 1910.179(j)(2)(iii), rigging hardware used during lifting services is required by OSHA to have a serial number or some other identifier. The most effective method of ensuring that the product you are using is as reliable as possible is to purchase components supplied by companies who maintain consistent and adequate quality. The company should clearly mark its components and finished products with the company name or logo, the component size or working load limit, and a code that is actively used by the manufacturer to control material and processes.

A forged-in identification code should be used to record the material grade and origin. This record should trace the material to the heat lot of material as rolled at the supplying mill. Verification checks of all materials purchased for forging must be done to ensure that the steel supplied meets the specifications required. This verification should be traceable by a forged-in product identification code. In brief, the source and verification of material actually used in forging must be able to be determined through appropriate documentation.

The permanent identification code should be used to maintain a record of which manufacturing facility produced the product as well as the approximate production dates. All quality records should reference the product identification code so that a history can be maintained. All product performance testing for audit and engineering purposes should also reference the product identification code.

Detailed performance, application, and warning information will assist in the proper use of products. This information is most effective when provided in supporting brochures and engineering information. An identification marking must be used to reference this information by use of a cross reference between the product code and the literature.

Proper performance data should include each item's working load limit, proof load and design factor. It should also include the item's manufacturing processes, such as heat treatment and galvanizing, and list any specification the product meets or exceeds. [1]

Hook Frequent Inspection Interval

The operator or an employer-designated person must conduct a daily visual inspection before use to identify any of the following in crane hooks:

- Cracks, nicks, gouges
- Deformation
- Damage from chemicals
- Damage or malfunction of the throat latch, when provided.

A visual examination by the operator (records not required) or the employer designated person is required for frequent inspections of hooks at the following intervals:

- Normal service: monthly

- Heavy service: weekly to monthly

- Severe service: daily to weekly.

Hook service classifications are determined by the following:

- Normal service involves operating at less than 85% of rated load except for isolated instances.

- Heavy service that involves operating at 85% to 100% of rated load as a regular specified procedure.

- Severe service is heavy service coupled with abnormal operating conditions.

Hook Periodic Inspection Interval

A qualified inspector is required to perform visual inspections and record any apparent external conditions to provide the basis for continuing evaluation. The periodic inspection intervals are:

- Normal service: yearly, with equipment in place.

- Heavy service: semiannually, with equipment in place unless external conditions indicate that disassembly should be done to permit detailed inspection.

- Severe service: quarterly, as in heavy service, except that detailed inspection may show the need for nondestructive type testing.

Hoist Chains

1910.179(1)(3)(iii)(a)

"Hoist chains, including end connections, for excessive wear, twist, distorted links interfering with proper function, or stretch beyond manufacturer's recommendations. Visual inspection daily; monthly inspection with a certification record which includes the date of inspection, the signature of the person who performed the inspection, and an identifier of the chain which was inspected."

Hoist Chains Frequent Inspection

Frequent inspection must include observations during operation. The determination must be made by a designated person whether any conditions found during the inspection constitute a hazard. If so, a more detailed inspection would be required.

During the inspection, hooks should be scrutinized for the following:

- Distortion, such as bending, twisting, or increased throat opening

- Wear

- Cracks, nicks, or gouges

- Latch engagement, damaged, or malfunctioning latch (if provided)

- Hook attachment and securing means.

Hoist Chains Periodic Inspection

Inspection of hooks shall be performed as defined in the frequent inspection section. Hooks having any of the following conditions shall be removed from service until repaired or replaced.

- Deformation. Any bending or twisting exceeding 10 degrees from the plane of the unbent hook.

- Throat Opening. Any distortion causing an increase in throat opening exceeding 15 percent.

- Wear. Any wear exceeding 10 percent of the original section dimension of the hook or its load pin.

Hoist Chains Inspection Records

Inspection records must include the date of inspection, the signature of the person who performed the inspection, and the serial number, or other identifier, of the hook inspected. Inspection records should be retained in the equipment history file. Following is a summary of inspection record requirements.

- Initial Inspection: A record of the initial inspection shall be made.

- Pre-use and Frequent Inspection: No records are required.

- Periodic Inspection: A record of periodic inspections shall be made.

OVERHEAD CRANES

Rated Load Markings/Tests

OSHA 1910.179(b)(5) requires that rated load markings be visibly marked on each side of overhead cranes with the following statement:

"Rated load marking. The rated load of the crane shall be plainly marked on each side of the crane, and if the crane has more than one hoisting unit, each hoist shall have its rated load marked on it or its load block and this marking shall be clearly legible from the ground or floor."

OSHA 1910.179 and OSHA 1910.180 respectively require all new, reinstalled, altered, repaired, and modified overhead and mobile cranes to be load-tested prior to initial use. Test loads shall not be less than 100 percent of the rated load of the crane or more than 125 percent of the rated load unless otherwise recommended by the manufacturer. Test reports must be placed on file where readily available to appointed personnel, show the test procedures used, and confirm the adequacy of repairs or alterations. Any load-testing of altered, repaired, and modified cranes may be limited to the functions affected by the alteration, repair, or modification, as determined by the designated person. While no standard requires load-test requirements after the replacement of load chain and rope, an operational test of the hoist must be made prior to returning the crane to service.

Maintenance Requirements

Only designated personnel are allowed to perform any required maintenance and repairs. If any deteriorated components or unsafe conditions are detected during the required inspections, they must be completed before the crane is allowed to be used. The requirements of 29 CFR 1910.147, The Control of Hazardous Energy (lockout/tagout), should be used to de-energize the crane.

Supervisor / Foreman

A supervisor/foreman or other employer-components shall ensure responsible inspectors, maintenance, and test personnel will have access to applicable adequate information as follows:

Operating instructions furnished by the manufacturer or the responsible maintenance/engineering organization including:

- Maintenance, repair, and parts information

- Manufacturer's recommendations as to points and frequency of lubrication, maintenance of lubrication levels, and types of lubricant to be used

- Maintenance or repair procedures from the manufacturer or from a responsible maintenance/engineering organization

- Wiring diagrams. Inspection, maintenance, and repair activities are documented in accordance with the requirements of this manual.

All cranes and derricks must be inspected, maintained, and repaired by qualified maintenance and repair personnel having the necessary tools to safely accomplish their work. For all equipment furnished with a manual, personnel responsible for inspection or maintenance must be familiar with the applicable contents of the manual.

Preventive Maintenance

The crane manufacturer's recommendations should be used to implement a preventive maintenance program. The employer must comply with the manufacturer's specifications and limitations applicable to the operation of any and all cranes and derricks. Where manufacturer's specifications are not available, the limitations assigned to the equipment shall be based on the determinations of a qualified engineer competent in this field, and such determinations will be appropriately documented and recorded.

> *Note: When maintenance is being performed on overhead cranes, "warning" or "out of order" signs must be placed on the crane. If there is personnel access underneath the crane, visible warning signs should be placed on the floor. Current lockout/tagout procedures must be followed.*

Maintenance Precautions

Before adjustments or repairs are started on overhead cranes, the following precautions shall be taken as applicable.

- The crane to be repaired shall be moved to a location where it will cause the least interference with other equipment and operations in the area.

- Controllers shall be set in the off condition.

- The main switch (crane disconnect) shall be de-energized and locked, tagged, or flagged in the de-energized position.

- When maintenance work creates a hazardous area on the floor beneath the crane or crane runway, effective markings and barriers need to be in place.

- Where other cranes are in operation on the same runway, rail stops or other means have to be provided to prevent interference with the idle crane or work area.

- If temporary protective rail stops or other means are not available or practical, a signal person would have to be placed at a visual vantage point for observing the approach of an active crane and warning its operator when reaching the limit of safe distance from the idle crane or work area.

- Only trained personnel can work on energized equipment when required adjustments and tests are being made.

- After maintenance work is completed and before restoring the crane to normal operation, the following activities need to be completed:

  1. Guards reinstalled

  2. Safety devices reactivated

  3. Replaced parts and loose material removed

  4. Maintenance equipment removed.

CRANE OPERATING PRACTICES

Load Limits

No crane should be loaded beyond its rated load except for test purposes, and provisions for special overrated load lifts are outlined in ASME B30 standards. However, no such lift should be made without the authorization of the crane manufacturer.

Handling the Load

The load must always be attached to the hook by means of slings or other devices of sufficient capacity. At no time is it permissible to wrap the hoist line around the load. Unless the crane is equipped with automatic drum and boom braking systems, the drum and boom brakes must be set to hold the load if the load is to remain suspended for any considerable length of time.

No crane should be loaded beyond the specifications of the load rating chart, except for test purposes. Total load always includes the lifted item and the rigging. Additionally, the block and hook may also be considered part of the load. Attachments to the boom such as a jib or auxiliary whip lines affect crane stability and may be considered part of the load as well as the crane hook, block, and load line.

When the precise load weight is not known, the person responsible for the lift shall ascertain that the weight does not exceed the crane rating at the radius at which the load is to be lifted. Any time the weight of the load is unknown and it is estimated to be potentially near the crane's capacity, a load-indicating device should be used.

Load Rating Charts

There have been a significant number of fatalities and injuries due to improper crane use. The use of appropriate documented capacity charts, crane configuration limitations, and hand signal illustrations are required to ensure the employee a safe place to work during the use of a crane or derrick. Without the required documentation to guide the user, cranes and derricks would be misused because information necessary for safe use would be unavailable. This documentation is used by employers and employees to determine the machine's capacity to lift and place loads and as a reference for proper hand signals to be used.

A manufacturer-provided, durable, legible rating chart(s) must be provided in the cab of each crane. Should the original chart(s) become illegible, a new chart(s) must be obtained from the crane manufacturer or designee. The crane's serial number should be printed or stamped on the chart by the manufacturer or designee. It is the responsibility of the employer to ensure that the proper load chart(s) is available in the cab and is in a location accessible to the operator while the operator is at the controls.

The rating chart(s) must include a full and complete range of manufacturer's crane load ratings at all stated operating radii, boom angles, work areas, boom lengths and configurations, jib lengths and angles (or offset), as well as alternative ratings for use and non-use of optional equipment on the crane, such as outriggers and extra counterweight, which affect ratings. Where ratings are limited by structural, hydraulic, or factors other than stability, the limitations shall be shown and emphasized on the rating charts.

The manufacturer should and normally will provide a work area chart(s) showing the limiting position of any load within areas indicated (e.g., over side, over rear, over front). Other information included will be a work area chart for which capacities are listed in the load rating chart. Work area setups derived from ASME B30.5 must be defined by diagrams supplied by the crane manufacturer.

When working at boom lengths or radii between the figures shown on the load capacity chart, the next lower capacity rating should be used. No lift should be performed until the capacity for boom lengths or radii between those listed on the rating page has been established.

Some operators still observe the crane's wheels coming off the ground to warn of an overload. This should not be permitted because cranes may suddenly tip over or the boom may collapse if the load is too heavy.

It is the responsibility of the employer to ensure that operators always operate within the cranes rated capacity and working radius. When required, the load capacity must be reduced until it is agreed upon by the crew that the load can be safety lifted.

Load Considerations

- Load radius: the horizontal distance between the center of the crane rotation to center of the load

- Boom length: including the jib, swing-away extension or any other attachments that may increase length of the boom

- Parts of line

- Quadrant of operations: the area of operation that the lift is being made in; note different quadrants usually have lower lifting capacities

- Boom angle: the angle formed between the horizontal plane of rotation and center line of the boom

- Weight of any attachments: jib, lattice extension, or auxiliary boom point

- Weight of handling devices: ball, block, and/or any necessary rigging.

Hand Signals

Hand signals to crane and derrick operators shall be those prescribed by the applicable ANSI standard for the type of crane in use. An illustration of the signals as shown in Figure 2.1 shall be posted at the job site.

Figure 2.1. Hand Signals for Crane Operators

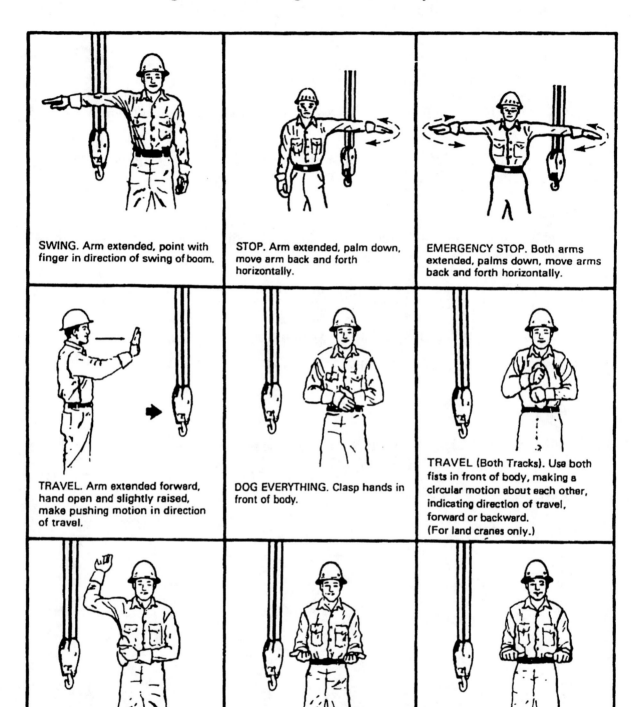

SWING. Arm extended, point with finger in direction of swing of boom.

STOP. Arm extended, palm down, move arm back and forth horizontally.

EMERGENCY STOP. Both arms extended, palms down, move arms back and forth horizontally.

TRAVEL. Arm extended forward, hand open and slightly raised, make pushing motion in direction of travel.

DOG EVERYTHING. Clasp hands in front of body.

TRAVEL (Both Tracks). Use both fists in front of body, making a circular motion about each other, indicating direction of travel, forward or backward. (For land cranes only.)

TRAVEL. (One Track) Lock the track on side indicated by raised fist. Travel opposite track in direction indicated by circular motion of other fist, rotated vertically in front of body. (For land cranes only.)

EXTEND BOOM (Telescoping Booms). Both fists in front of body with thumbs pointing outward.

RETRACT BOOM (Telescoping Booms). Both fists in front of body with thumbs pointing toward each other.

Figure 2.1. Hand Signals for Crane Operators (continued)

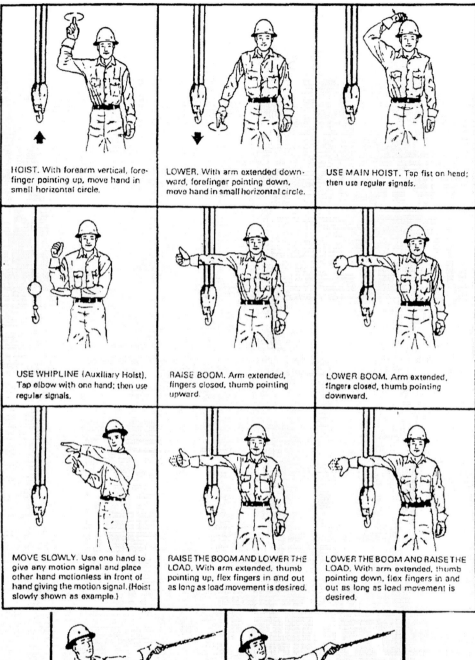

HOIST. With forearm vertical, fore-finger pointing up, move hand in small horizontal circle.

LOWER. With arm extended downward, forefinger pointing down, move hand in small horizontal circle.

USE MAIN HOIST. Tap fist on head; then use regular signals.

USE WHIPLINE (Auxiliary Hoist). Tap elbow with one hand; then use regular signals.

RAISE BOOM. Arm extended, fingers closed, thumb pointing upward.

LOWER BOOM. Arm extended, fingers closed, thumb pointing downward.

MOVE SLOWLY. Use one hand to give any motion signal and place other hand motionless in front of hand giving the motion signal. (Hoist slowly shown as example.)

RAISE THE BOOM AND LOWER THE LOAD. With arm extended, thumb pointing up, flex fingers in and out as long as load movement is desired.

LOWER THE BOOM AND RAISE THE LOAD. With arm extended, thumb pointing down, flex fingers in and out as long as load movement is desired.

EXTEND BOOM (Telescoping Boom). One Hand Signal. One fist in front of chest with thumb tapping chest.

RETRACT BOOM (Telescoping Boom). One Hand Signal. One fist in front of chest, thumb pointing outward and heel of fist tapping chest.

SUMMARY

The need for identifying responsibilities extends from the employer, to the crane operator, to the employee(s) doing rigging for the load to be lifted. Everyone must be aware of the serious nature of the hazards that are involved. This is due to the following: the magnitude of the risk, the high rate of fatalities and serious injuries relative to the number of workers exposed, the potential for catastrophic consequences due to accidents, and the considerable knowledge available about effective protective measures.

Employers are responsible for the operation, inspection, maintenance, and repair of equipment and components for ensuring all hoisting equipment is operated in a safe manner and only operated by qualified personnel. Employers are also responsible for work assignments and must ensure that such assignments do not exceed personnel qualifications.

Employees assigned to operate cranes must be trained and qualified to the level of proficiency consistent with assigned duties. A crane operator is responsible for those operations under the operator's direct control, and if there is ever any doubt as to safety, it is the responsibility of the operator to consult with the foreman or supervisor before continuing rigging or lifting the loads.

As with the crane operator, only qualified and authorized employees should be permitted to perform rigging functions. The prescribed responsibilities of those employees performing such rigging assignments must be identified. One identified responsibility of the rigger would be to ensure that the rigging equipment and materials have the required capacity for the job and that all items are in good condition, are currently qualified (inspection is up-to-date), and are properly used.

Inspections of the crane and associated rigging equipment are required to be performed at intervals depending upon its activity, severity of service, and environment, or as specifically indicated by OSHA or ANSI/ASME standards. One requirement by OSHA regulations dictates that such equipment be inspected during initial use, frequently, and annually thereafter by a "competent person."

REFERENCES

1. The Crosby Group Incorporated. *General Catalog*, Tulsa, OK. 1996. pp. 2–22.

2. The Crosby Group Incorporated. *User's Lifting Guide*, Tulsa, OK. pp. 6.

3

HOISTING AND RIGGING OPERATIONS

All hoisting and rigging operations consist of three components: the crane that is used, the crane operator, and the rigging crew that is responsible for rigging the load and hooking it to the crane. All components must work in unison in order to accomplish the intended task. As mentioned earlier, a crane requires varying inspections depending upon its usage. From the initial inspection to the frequent and periodic, all crane inspections must be performed at the intervals defined by OSHA. Without proper inspection, deficiencies may exist that contain safety hazards that can lead to incidents involving personnel injury or property damage. All inspection, maintenance, and repair activities are documented in accordance with the requirements of all current rules and regulations.

Once the operator has performed the pre-use inspection and is assured all other inspections have been performed, it is contingent on him or her to safely operate the equipment. This is done by following all equipment operating guidelines; and, for mobile cranes, the load charts must be consistently used. The operator must abide by any restrictions placed on the use of the equipment and ensure that the load will not exceed the rated capacity of the equipment. By working closely with the rigging crew, the operator can be assured that all rigging equipment has been inspected, correct slings and hardware have been selected, and the load is properly attached.

There are several beneficial aids available from different companies to support the operator and the rigging crew in making the correct choices when preparing to make a lift. An example of such an aid is the Crosby® User's Lifting Guide [1] used in different areas of this text. The lifting guide is comprised of the following 12 tables, figures, and charts:

- Risk Management, Terminology, and Inspection of Fittings

- Wire Rope Sling Facts

- Wire Rope Sling Capacities

- Chain Sling Capacities

- Web Sling Capacities

- The Basic Rigging Plan/Basic sling Operating Practices B30.9

- Sling Angles

- Load Distribution – Rigging

- Rigging Hardware – Shackles/Hooks

- Rigging Hardware – Wire Rope Clips/Turnbuckles

- Rigging Hardware – Rigging Hardware

- Center of Gravity/Calculating Weight.

Table 3.1 is the first of the above-mentioned twelve tables and provides an overview of the elements comprised in a hoisting and rigging program. This includes the definition of risk management, commonly used terminology, and information on the inspection of fittings.

LOAD CONTROL

Estimating Load Weights

Before any lift is made, the operator must know the weight of the load to be lifted or use a crane equipped with a load-indicating device that he or she has been trained to use. The weight of the load can be determined by the operator or the individual accountable for the project. If the operator is ever doubtful as to the weight of load, the weight of the load should be determined from shipping papers, manufacturer's information, catalogs, or blueprints. If a load has been shipped by truck, often the truck will have been weighed at several different scales en route to the jobsite.

Table 3.1. Lifting Definitions/Terminology [1]

RISK MANAGEMENT

DEFINITION

COMPREHENSIVE SET OF ACTIONS THAT REDUCES THE RISK OF A PROBLEM, A FAILURE, AN ACCIDENT

YOU NEED

- PRODUCT KNOWLEDGE
- APPLICATION KNOWLEDGE
- MANUFACTURER OF KNOWN CAPABILITY
- PRODUCTS THAT ARE CLEARLY IDENTIFIED WITH THE FOLLOWING;
  1. MANUFACTURER'S NAME AND LOGO
  2. LOAD RATING OR SIZE THAT REFERENCES RATINGS
  3. TRACEABILITY CODE

A GOOD RISK MANAGEMENT PROGRAM RECOGNIZES

- PERFORMANCE REQUIREMENTS INCLUDE THE FOLLOWING:
  1. LOAD RATED PRODUCTS
  2. QUENCH AND TEMPERED
  3. ABILITY TO DEFORM WHEN OVERLOADED.
  4. ABILITY TO WITHSTAND REAL WORLD LOADING IN DAY TO DAY USE, TOUGHNESS.

TERMINOLOGY

WORKING LOAD LIMIT (WLL)

THE MAXIMUM MASS OR FORCE WHICH THE PRODUCT IS AUTHORIZED TO SUPPORT IN A PARTICULAR SERVICE.

PROOF TEST

A TEST APPLIED TO A PRODUCT SOLELY TO DETERMINE INJURIOUS MATERIAL OR MANUFACTURING DEFECTS.

ULTIMATE STRENGTH

THE AVERAGE LOAD OR FORCE AT WHICH THE PRODUCT FAILS OR NO LONGER SUPPORTS THE LOAD.

DESIGN FACTOR

AN INDUSTRIAL TERM DENOTING A PRODUCT'S THEORETICAL RESERVE CAPABILITY; USUALLY COMPUTED BY DIVIDING THE CATALOG ULTIMATE LOAD BY THE WORKING LOAD LIMIT. GENERALLY EXPRESSED AS A RATIO, e.g. 5 TO 1.

INSPECTION OF FITTINGS

DEFORMATION

CROSBY RECOMMENDS THAT NO SIGNIFICANT DEFORMATION BE ALLOWED.

WEAR

ACCEPTABLE LIMITS:
5%   WEAR IN THE THROAT & EYE OF HOOKS AND OTHER CRITICAL SECTIONS OF ALL FITTINGS.
10%   WEAR IN OTHER AREAS.

CRACKS

REMOVE FITTINGS FROM SERVICE WITH CRACKS.

WELDING AND MODIFICATIONS

DO NOT WELD ON OR MODIFY FITTINGS OR BLOCKS.

FOR ADDITIONAL SUPPORT

P.O. BOX 3128
TULSA OKLAHOMA 74101
(918)834-4611

A comparison of those truck weights with the original weight that was furnished for the load can possibly confirm the weight of the load.

Other methods the operator can use to calculate the load weight include:

- Using a scale to weigh the load

- Reviewing associated paperwork such as shipping documents, drawings, or information furnished by the manufacturer

- Calculating or estimating the surface area of the shape of the material to be lifted and multiply that by the weight pounds per square foot of the material

- Calculating or estimating the weight based on the volume of the material to be lifted and multiply that by the weight pounds per cubic foot of the material.

Center of Gravity

The center of gravity is the location of the load where the weight of the load is evenly distributed at all attachment points. Some manufacturers label the center of gravity on the load while others place it in drawings or catalogs of the product. The balance point and center of gravity must be located directly below the hoisting hook before any load is lifted. If the load is not balanced when lifted, gravity will attempt to balance it. When the load is not centered below the hook, and two slings are used, one will assume a greater share of the load.

If the center of gravity is not known and cannot be accurately determined before starting the lift, the load should be connected and slightly lifted to check for stability. If the load appears unstable, it should be lowered and the lifting point adjusted. This process can then be repeated until the operator determines that the load is stable and can be lifted. Figure 3.1, "Center of Gravity and Sling Loading/Calculating Weights" and Figure 3.2, "Load Distribution – Rigging" provide guidance on calculating the center of gravity and load distribution.

Figure 3.1. Center of Gravity and Sling Loading/Calculating Weights [1]

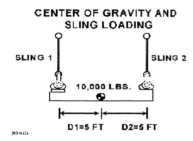

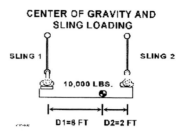

WHEN LIFTING VERTICALLY, THE LOAD WILL BE SHARED EQUALLY IF THE CENTER OF GRAVITY IS PLACED EQUALLY BETWEEN THE PICK POINTS. IF THE WEIGHT OF LOAD IS 10,000 LBS., THEN EACH SLING WILL HAVE A LOAD OF 5,000 LBS. AND EACH SHACKLE AND EYEBOLT WILL ALSO HAVE A LOAD

WHEN THE CENTER OF GRAVITY IS NOT EQUALLY SPACED BETWEEN THE PICK POINTS, THE SLINGS AND FITTINGS WILL NOT CARRY AN EQUAL SHARE OF THE LOAD. THE SLING CONNECTED TO THE PICK POINT CLOSEST TO THE CENTER OF GRAVITY WILL CARRY THE GREATEST SHARE OF THE LOAD.

SLING 2 IS CLOSEST TO COG. IT WILL HAVE THE GREATEST SHARE OF THE LOAD.

SLING 2 = 10,000 X 8 / (8+2) = 8,000 LBS.=
SLING 1 = 10,000 X 2 / (8+2) = 2,000 LBS.

CALCULATE WEIGHT EXAMPLE - FLATS

WEIGHT = W x L x t x UNIT WEIGHT

L= 5 FT.

t = 1"

W = 2 FT.

IF STEEL

CALCULATE WEIGHT EXAMPLE - SOLID CYLINDER

$$WEIGHT = \frac{3.14 \times D^2 \times L \times UNIT\ WEIGHT}{4}$$

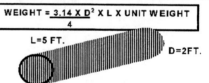

L=5 FT.

D=2 FT.

OF 5,000 LBS.

IF STEEL

    UNIT WEIGHT IS 490 LBS/FT³

    WEIGHT = 2' X 5' X 1/12' X 490 LBS/FT³ = 408 LBS.

IF ALUMINUM

    UNIT WEIGHT IS 165 LBS/FT³

    WEIGHT = 2' X 5' X 1/12' X 165 LBS/FT³ = 138 LBS.

IF CONCRETE

    UNIT WEIGHT IS 150 LBS/FT³

    WEIGHT = 2' X 5' X 1/12' X 150 LBS/FT³ = LBS.

IF STEEL

    UNIT WEIGHT IS 490 LBS/FT³

    WEIGHT = 3.14 (2' X 2') X 5' X 490 LBS/FT³ = 7693 LBS.

IF CONCRETE

    UNIT WEIGHT IS 150 LBS/FT³

    WEIGHT = 3.14 (2' X 2') X 5' X 150 LBS/FT³ = 2355 LBS.

Figure 3.2. Load Distribution - Rigging [1]

LOADWALKING

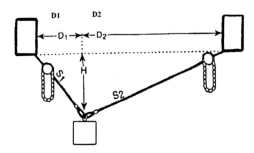

LOAD ON SLING CALCULATED

TENSION 1= LOAD X D2 X S1/H(D1 + D2)
TENSION 2= LOAD X D1 X S2/H(D1 + D2)

POSITIVE LOAD CONTROL

YES

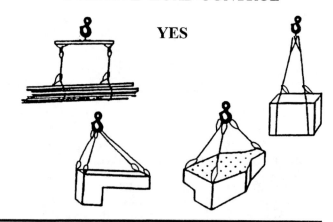

TYPES OF HITCH CONSIDERATION

LOAD CONTROL
THE ABILITY OF THE SLING TO CONTROL
THE MOVEMENT OF THE LOAD BEING
LIFTED

CAPACITY
THE LOAD CAPACITY OF THE SLING AND
TYPE OF HITCH

TYPE OF SLING
WIRE ROPE
CHAIN
WEBBING

CENTER OF GRAVITY
THE LOCATION OF THE CENTER OF THE
LOAD'S WEIGHT

TRIPLE AND QUAD LEG SLING

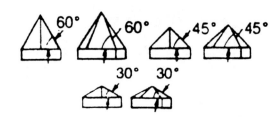

TRIPLE LEG SLINGS HAVE 50% MORE CAPACITY THAN
DOUBLE LEG ONLY IF THE CENTER OF GRAVITY IS IN
CENTER OF CONNECTION POINT AND LEGS ARE
ADJUSTED PROPERLY (EQUAL SHARE OF THE LOAD)

QUAD LEG SLINGS OFFER IMPROVED STABILITY BUT
DO NOT PROVIDE INCREASED LIFTING CAPACITY.

UNEQUAL LEGS

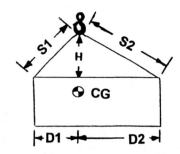

LOAD ON SLING CALCULATED

TENSION 1= LOAD X D2 X S1/H(D1 + D2)
TENSION 2= LOAD X D1 X S2/H(D1 + D2)

REEVING INCREASES LOADS

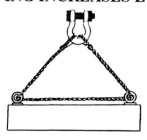

REEVING THROUGH CONNECTIONS TO LOAD
INCREASES LOAD ON CONNECTIONS FITTINGS
BY AS MUCH AS TWICE—

DO NOT REEVE!

WIRE ROPE

Design and Construction

Wire ropes are composed of independent parts—wires, strands, and cores—that continuously interact with one other during service. Wire rope engineers design those parts in differing steel grades, finishes and a variety of constructions to attain the best balance of strength, abrasion resistance, crush resistance, bending, fatigue resistance, and corrosion resistance for each application.

To select the best wire rope for each application, one must know the required performance characteristics for the job and enough about the wire rope design to select the optimum combination of wire rope properties.

The following information is presented as a basic guide on strand constructions. Wire rope companies have engineers and field service representatives available to provide more specific recommendations.

Strands are designed with various combinations of wires and wire sizes to produce the desired resistance to fatigue and abrasion. Generally, a small number of large wires will be more abrasion resistant and less fatigue resistant than a large number of small wires. Independent Wire Rope Core (IWRC) provides good crush resistance and increased strength, while Fiber Core provides excellent flexibility. Most wire ropes are made with round wires. Both triangular and shaped wires are also used for special constructions. Generally, the higher the strength of the wire, the lower its ductility will be. Table 3.2 describes the different types of wire rope strand constructions.

Table 3.2. Wire Rope Strand Construction

Single Size	The basic construction has wires of the same size wound around a center.
Seale	Large outer wires with the same number of smaller inner wires around a core wire. Provides excellent abrasion resistance but less fatigue resistance. When used with an IWRC, it offers excellent crush resistance over drums.
Filler Wire	Small wires fill spaces between large wires to produce crush resistance and a good balance of strength, flexibility, and resistance to abrasion.
Warrington	Outer layer of alternatively large and small wires provides good flexibility and strength but low abrasion and crush resistance.
Seale Filler Wire Filler Wire Seale Warrington Seale Seale Warrington Seale	Many commonly used wire ropes use combinations of these basic constructions. D/d ratio is the ratio of the diameter around which the sling is bent
Multiple Operation	One of the above strand designs may be covered with one or more layers of uniform-sized wires.

Finish

Bright finish is suitable for most applications. Galvanized finish is available for corrosive environments. Plastic jacketing is also available on some constructions.

Grade and Construction

Grade and construction of wire rope slings is normally bright Improved Plow Steel, Extra Improved Plow Steel, or Extra Extra Improved Plow Steel, 6x19 or 6x36 classification, right regular lay. Stainless steels and other special grades are available for special applications. IWRC rope has a higher rated capacity than fiber core rope for mechanically spliced slings, but the same rated capacity for hand-tucked slings. This is because when making a hand-tucked splice, the core (IWRC) of the rope is cut in the splice area and does not add to the overall strength of the sling. Rated capacities of slings using galvanized rope may be less than the same grade bright rope sling. The sling manufacturer should be consulted regarding rated capacities for these types of slings. Figure 3.3, "Wire-Rope Lays," provides drawings and Table 3.3, "Wire Rope Description," provides a description of the assorted types of lays.

Figure 3.3. Wire-Rope Lays [1]

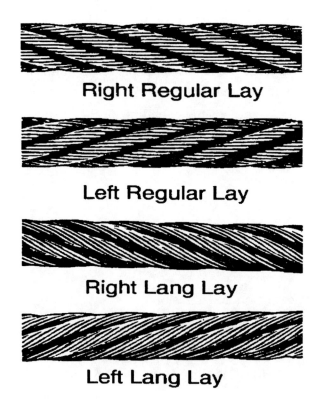

Right Regular Lay

Left Regular Lay

Right Lang Lay

Left Lang Lay

Table 3.3. Wire Rope Description

Lay	Definition	Characteristics
Regular Lay	Most common lay in which the wires wind in one direction and the strands the opposite direction.	Less likely to kink and untwist; easier to handle; more crush resistant than lang lay.
Lang Lay	Wires in strand and strands of rope wing the same direction.	Increased resistance to abrasion; greater flexibility and fatigue resistance than regular lay; will kink and untwist.
Right Lay	Strands wound to the right around the core.	The most common construction.
Left Lay	Strands wound to the left around the core.	Used in a few special situations ; cable tool drilling line, for example.

GUIDELINES FOR SELECTION

Strength

Wire rope must have the strength required to handle the maximum load plus the design factor. The design factor is the ratio of the breaking strength of the rope to the maximum working load. To establish the proper design factor, several operating characteristics should be considered:

- Speed of operation
- Acceleration and deceleration
- Length of rope
- Number, size, and location of sheaves and drums
- Rope attachments
- Conditions causing corrosion and abrasion
- Danger to human life and property.

Table 3.4. Wire Rope Design Factors

Type of Service	Minimum Factor
Guy Ropes	3.5
Overhead and Gantry Cranes	3.5
Derricks	3.5
Jib and Pillar Cranes	3.5
Miscellaneous Hoisting Equipment	5.0
Wire Rope Slings	5.0

Fatigue, Abrasion, and Crush Resistance

Smaller wires are the key to bending performance when wire ropes are subjected to repeated bending over sheaves or drums. The more outer wires for a given size wire rope, the better resistance to bending fatigue. Lang lay and large outer wires provide resistance to abrasion while an IWRC and large outer wires will provide best crush resistance. Fiber core, lang lay, and smaller wires provide a more flexible wire rope.

Installation, Operation, and Maintenance

The primary factor in wire rope performance is selecting a wire rope with the best combination of properties for the job. The service life of that rope can be greatly extended by following a planned program of installation, operation, maintenance, and inspection to avoid the following most common causes of wire rope failure.

- KINKING will result in permanent rope deformation and localized wear. It is generally caused by allowing a loop to form in a slack line and then pulling the loop down to a tight permanent set. This prevents them from sliding and adjusting, and reduces rope strength.

- OVERLOADING results in accelerated wear, abrasion, rope crushing, and distortion on drums and sheaves, and could result in complete rope failure.

- DRAGGING wire rope over a bank or some other object results in localized wear, which means shorter wear.

- IMPROPER SPOOLING results in crushed and distorted ropes and comes from careless installation of the rope.

- WHIPPING a line, which results in many squared off broken wires, comes from jerking or running the wires loose.

Unloading, Unreeling, and Uncoiling Wire Rope

Suitable precautions should be taken to prevent dropping of reels or coils during unloading and moving. If the reel should collapse, it may be impossible to remove the rope without serious damage.

Special care should be taken in unreeling wire rope to avoid kinking, which can result in permanent damage to the rope. The reel should be mounted on jacks or a turntable so that it will move more freely. It should be unreeled straight and under enough tension to keep it from starting a loop.

A coil should be unwound by rolling along the floor like a hoop. Coils should never be laid flat and the free end pulled out.

Proper practices for transferring rope from reel to drum include:

- Placing the reel as far away from the drum as possible in order to avoid putting any turn into the rope

- Winding the rope from top-to-top or bottom-to-bottom to avoid reverse bends, which tend to make a rope harder to handle

- Using enough tension to avoid kinking.

There is usually only one way to install rope on a grooved drum. On ungrooved drums, the "rule of thumb" guides installation. The fist represents the drum; the index finger the wire rope; and the thumb the direction of the proper dead-end location. Use the right hand for right lay ropes, the left hand for left lay ropes. For overwinding, the palm is down; for underwinding, the palm is up. Most drum anchors are set for right lay ropes since it the most common specification.

On installations where the rope passes over a sheave onto the drum, the maximum fleet angle (angle between the center line of the sheave and the rope) should not be more than 1½ degrees for a smooth-faced drum and 2 degrees for a grooved drum. Larger fleet angles may cause excessive wear from rubbing against the flanges of the sheaves as well as excessive crushing and abrasion of the rope on the drum.

Break in/Operation

A few trips through the working cycle at slow speed and light load will set the strands firmly in place for smooth, efficient operation. On applications using a wedge socket, such as drag and hoist ropes, it is also a good idea to cut off a short section of rope to allow twist to run out and to equalize the strands.

Skillful operation is important to wire rope performance. Rapid acceleration, shock loading, and excessive vibration can cause permanent rope failure. Smooth, steady application of power by the equipment operator can add significantly to wire rope service life.

Shifting Wear Points

Some sections of most wire ropes get more wear than others. A regular inspection program will identify points of wear and lead to wear-shift practices that will extend wire rope life. In many common situations, cutting off short lengths of the rope will redistribute the points of maximum wear:

- Rope on a drum with two or more layers will wear at the point where the rope starts each successive layer.
- Crane ropes will fatigue at an equalizer sheave. Careful inspection is required to identify fatigue points.
- Hoist ropes will frequently fail from vibration fatigue at socket clips and dead end points.

On most installations, wear and fatigue are more severe on one half of the rope than the other. Changing a rope end-for-end more evenly distributes wear and fatigue from repeated bending and vibration. To determine the diameter of a wire rope (the diameter of the smallest circle that will enclose all the strands) measurements should be made to the outer wires.

Wire Rope Lubrication

Factory lubrication is not always sufficient to last the useful life of wire rope. Periodic field lubrication may be required to minimize friction and provide corrosion protection. The following are important guides for field lubrication:

- Ropes should be inspected frequently to determine the need for lubrication.

- Clean the rope throughly with a wire brush, scraper, or compressed air to remove foreign material and old lubricant from the valleys between the strands and the spaces between the outer wires.

- The lubricant should be applied at a point where the rope is being bent in order to promote penetration within the strands. It may be applied by pouring, dripping, or brushing.

- Used motor oil is not recommended as a wire lubricant.

- Any lubricant used should be light bodied enough to penetrate the rope and it should also contain a corrosion inhibitor.

In the sling industry, similar terms are often applied to designate Rated Capacity. The term *Rated Load* is commonly used to describe Rated Capacity. Another term, *Working Load Limit (WLL)*, is often used to describe Rated Capacity. The WLL term, however, is used much more commonly in alloy chain slings and is not common when referring to the Rated Capacity of wire rope slings.

Matching Grooves to the Wire Rope

Grooves should be spaced so that one wrap of rope does not rub against the next wrap during operation. Grooves in drums and sheaves should be slightly larger than the wire rope to permit the rope to adjust itself to the groove. Tight grooves will cause excessive wear to outer wires; large grooves do not support the rope properly.

Wire ropes are manufactured slightly larger than nominal size. Maximum allowable oversize tolerances are shown in Table 3.5.

Table 3.5. Maximum Allowable Oversize Tolerances

Nominal Rope Diameter (Inches)	Tolerance	
	Under	Over
up to 1/8	- 0	0.08
Over 1/8 to 3/16	- 0	0.07
Over 3/16 to 1/4	- 0	0.06
Over 1/4	- 0	0.05

As a rope is run through a groove, both the rope and the groove become smaller. A used groove can be too small for a new rope, thus accelerating rope wear. A compromise between rope life and matching frequency must be made.

Grooves should have an arc of contact with the wire rope between 135 and 150 degrees. They should be tapered to permit the rope to enter and leave the field smoothly. Field inspection groove gauges are made to nominal diameter of the rope plus ½ of the allowable rope oversize tolerance. When the field inspection gauge fits perfectly, the groove is at the minimum possible contour.

Calculating Drum Capacity

The length of rope that can be wound on a drum or reel may be calculated as shown in Table 3.6.

Table 3.7 provides values of K for twenty different rope diameters.

Table 3.6. Calculating Drum Capacity

The length of rope that can be wound on a drum or reel may be calculated as follows:

L = the length of rope in feet. All other dimensions are in inches.

L = (A + D) x A x B x K

K = Constant obtained by dividing 0.2618 by the square of the actual rope diameter.

A = $\frac{H - D}{2}$ — desired clearance (in)

B = Traverse in inches

D = Barrel diameter in inches

H = Flange diameter in inches

L = Rope length in feet

Table 3.7. Values of Constant (K)

Rope Dia.	K	Rope Dia.	K
1/4"	3.29	1 1/8'	0.191
5/16"	2.21	1 1/4"	0.152
3/8"	1.58	1 3/8"	0.127
7/16"	1.19	1 2"	0.107
1/2"	.925	1 5/8"	0.0886
9/16"	.741	1 3/4"	0.0770
5/8	.607	1 7/8"	0.0675
3/4"	.428	2"	0.0597
7/8	.308	2 1/8"	0.0532
1"	.239	2 1/4"	0.0476

INSPECTION OF WIRE ROPE AND STRUCTURAL STRAND

Carefully conducted inspections are necessary to ascertain the condition of wire rope at various stages of its useful life. The object of wire rope inspection is to allow for removal of the rope from service before the rope's condition, as a result of usage, could pose a hazard to continued normal operations.

The individual making the inspection should be familiar with the product and the operation because his/her judgment is a most critical factor. Various safety codes, regulations, and publications give inspection requirements for specific applications. The following inspection procedure serves as a model of typical inspection requirements.

Frequent Inspection

All running ropes and slings in service should be visually inspected once each working day. A visual inspection consists of observation of all rope and end connections that can reasonably be expected to be in use during daily operations. These visual observations should be concerned with discovering gross damage such as listed below, which may be an immediate hazard:

- Distortion of the rope such as kinking, crushing, unstranding, birdcaging, main strand displacement, or core protrusion

- General corrosion

- Broken or cut strands

- Number, distribution, and type of visible broken wires

- Lubrication: Special care should be taken when inspecting portions subjected to rapid deterioration such as flange points, crossover points, and repetitive pickup points on drums.

Special care should also be taken when inspecting certain ropes such as:

- Rotation-resistant ropes such as 19x7 and 8x19, because of their higher susceptibility to damage and increased deterioration when working on equipment with limited design parameters

- Boom hoist ropes, because of the difficulties of inspection and important nature of these ropes

- When damage is discovered, the rope should either be removed from service or given an inspection as detailed in the next section.

Periodic Inspection

The inspection frequency should be determined by a qualified person and should be based on such factors as:

- Expected rope life as determined by experience on the particular installation or similar installations

- Severity of environment

- Percentage of capacity lifts

- Frequency rates of operation

- Exposure to shock loads.

Periodic inspections with a signed report should be performed by an appointed or authorized person. This inspection should cover the *entire length* of rope. The individual wires in the strands of the rope should be visible to this person during the inspection. Any deterioration resulting in appreciable loss of original strength, such as described below, should be noted and determination made as to whether further use of the rope would constitute a hazard:

- Distortion of the rope such as kinking, birdcaging, crushing, unstranding, main strand displacement, or core protrusion

- Reduction of rope diameter below normal diameter due to loss of core support, internal or external corrosion, or wear of outside wires

- Severely corroded or broken wires at end connections

- Severely corroded, cracked, bent, worn, or improperly applied end connections

- Lubrication.

    Special care should be taken when inspecting portions subjected to rapid deterioration such as the following:

    - Portions in contact with saddles, equalizer sheaves, or other sheaves where rope travel is limited

    - Portions of the rope at or near terminal ends where corroded or broken wires may protrude.

Rope Replacement

No precise rules can be given for determination of the exact time for replacement of rope, since many variable factors are involved. Continued use in this respect depends largely upon good judgment by an appointed or authorized person in evaluating remaining strength in a used rope, after allowance for deterioration disclosed by inspection. Continued rope operation depends upon this remaining strength.

Conditions such as the following should be sufficient reason for questioning continued use of the rope or increasing the frequency of inspection:

- In running rope, six randomly distributed broken wires in one lay or three broken wires in one strand in one lay. (The number of wire breaks beyond which concern should be shown varies with rope usage and construction.) For general application, six and three are satisfactory. Ropes used on overhead and gantry cranes (as defined in ASME B-30, 2-1983) can be inspected to four and two. Wire rope removal criteria are based on the use of steel sheaves. If synthetic sheaves are used, consult the sheave or equipment manufacturer.

- One outer wire broken at the contact point with the core of the rope which has worked its way out of the rope structure and protrudes or loops out from the rope structure

- Wear of one-third the original diameter of outside individual wires

- Evidence of any heat damage from any cause

- Valley breaks

- In standing ropes, more than two broken wires in one lay in section beyond end connections or more than one broken wire at end connection

- Reductions from nominal rope diameter (Table 3.8).

Replacement rope shall have a strength rating as great as the original rope furnished by the equipment manufacturer or as originally specified. Any deviation from the original size, grade, or construction shall be specified by the equipment manufacturer, original design engineer, or a qualified person.

Ropes Not In Regular Use

All rope that has been idle for a period of a month or more due to shutdown or storage of equipment on which it is installed should be given inspections as previously described before being placed back in service. This inspection should be for all types of deterioration and should be performed by an appointed or authorized person.

Table 3.8. Reductions From Nominal Rope Diameter

Reduction of	Nominal Rope Diameters
1/64"	up to 5/16"
1/32"	5/16" thru 1/2"
3/64"	1/2 thru 3/4"

Frequent Inspection Records

No records required.

Periodic Inspection Records

In order to establish data as a basis for judging the proper time for replacement, a signed report of rope condition at each periodic inspection should be kept on file. This report should include points of deterioration previously described.

A long range inspection program should be established and should include records of examination of ropes removed from service so a relationship can be established between visual observation and actual condition of the internal structure.

Wire Rope Slings

Design Factor

Design factor is a number that is divided into the nominal strength of a sling to arrive at a rated capacity. A design factor is necessary to allow for conditions such as wear, abrasion, damage, and variations in load that are not readily apparent. Design factors have been established that allow the sling to give the most efficient service to the user. Rated capacity tables contained in this text are based on a design factor of five (5). Other design factors may be applied for engineered lifts; however, the sling manufacturer should always be consulted.

Sling rated capacity is based upon the minimum breaking force, formerly called nominal (catalog) strength, of the wire rope used in the sling and other factors that affect the overall strength of the sling. These other factors include splicing efficiency, number of parts of rope in the sling, type of hitch (e.g., straight pull, choker hitch, basket hitch, etc.), diameter around which the body of the sling is bent (D/d), and the diameter of pin used in the eye of the sling.

D/d Ratio

D/d ratio is the ratio of the diameter around which the sling is bent divided by the body diameter of a single part sling, or the component rope diameter in a multi-part sling. This ratio has an effect on the rated capacity of the sling only when the sling is used in basket hitch. Tests have shown that whenever wire rope is bent around a diameter, the strength of the rope is decreased. Figure 3.4 illustrates the percentage of decrease to be expected. This D/d ratio is applied to wire rope slings to ensure that the strength in the body of the sling is at least equal to the splice efficiency.

Figure 3.4. D/d Ratio [2]

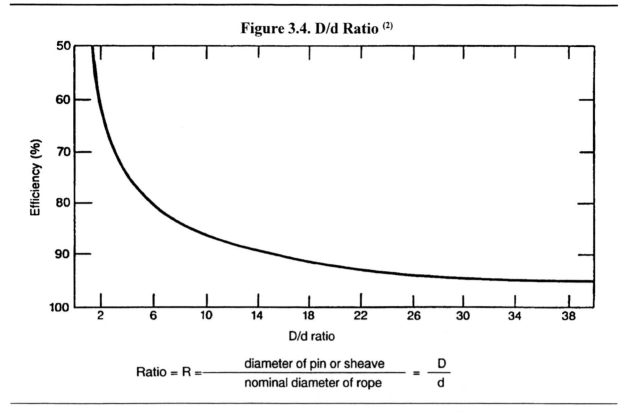

$$\text{Ratio} = R = \frac{\text{diameter of pin or sheave}}{\text{nominal diameter of rope}} = \frac{D}{d}$$

When D/d ratios smaller than those listed in the rated capacity tables are necessary, the rated capacity of the sling must be decreased.

The ratio of the diameter of the wire rope to the diameter of operating sheaves and drums (D/d) is particularly important to service life. A sheave or drum that is too small for the rope diameter will cause premature failure due to bending stress. Efficiency falls as the D/d ratio becomes smaller. This curve, based on static test data only, illustrates the decline of bending efficiency for 6x19 and 6x37 classification ropes as the D/d ratio is reduced.

Rated Capacity

Rated capacity is the maximum static load a sling is designed to lift using new, unused rope. The tables in this text give rated capacities in tons. Rated capacities contained in all the tables were calculated using component rope strength as a basis. Due to rounding of numeric values, rated capacity values for 2, 3, or 4 leg slings may not be even multiples of single leg values. Rounding also accounts for small differences in values between tables in various publications, and these small differences should not be construed to be in error.

Choker Hitch

Choker hitch configurations affect the rated capacity of a sling. This is because the sling or body is passed around the load, through one end attachment or eye, and is suspended by the other end attachment or eye. The contact of the sling body with the end attachment or eye causes reduction of sling efficiency at this point. If a load is hanging free, the normal choke angle is approximately 135 degrees. When the choke angle is less than 135 degrees, an adjustment in the choker rated capacity must be made (Figure 3.5). Extreme care should be taken to determine the choke angle as accurately as possible because as choke angles (degrees) increase, the decrease in rated capacity is dramatic. Slings with choke angle greater than 135 degrees are unstable and are not recommended.

Sling Angle

Sling angle is the angle measured between a horizontal plane and the sling leg or body. This angle is very important and can have a dramatic effect on the rated capacity of the sling. As shown in Figure 3.5., when this angle decreases, the load on each leg increases. This principle applies whether one or more slings are used to pull at an angle. Sling angles of less than 30 degrees shall not be used.

Nominal Splice Efficiency

Nominal splice efficiency is the efficiency of the sling splice. Any time wire rope is disturbed such as in splicing an eye, the strength of a rope is reduced. This reduction must be taken into account when determining the nominal sling strength and in calculating the rated capacity. Each type of splice has a different efficiency, thus the difference in rated capacities for different types of slings. Nominal splice efficiencies have been established after many hundreds of tests over years of testing.

Figure 3.5. Sling Angles [1]

TWO LEGGED SLING – WIRE ROPE, CHAIN, SYNTHETICS

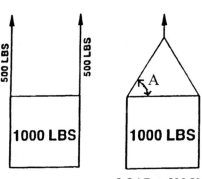

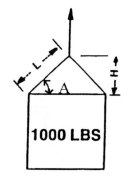

HORIZONTAL SLING ANGLE (A) DEGREES	LOAD ANGLE FACTOR = L/H
90	1.000
60	1.155
50	1.305
45	1.414
30	2.000

LOAD = 500 X
LOAD ANGLE FACTOR

LOAD IN EACH SLING
= L/H X 500

LOAD ON EACH LEG OF SLING
= VERTICAL LOAD X
LOAD ANGLE FACTOR

A = HORIZONTAL SLING ANGLE

CHOKER HITCHES WIRE ROPE

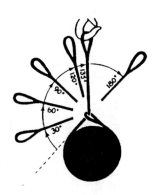

ANGLES OF CHOKE

120 - 180
90 - 119
60 - 89
30 - 59

SLING RATED LOAD
PERCENTAGE OF SINGLE
LEG SLING CAPACITY

75%
65%
55%
40%

CHOKER HITCHES WIRE ROPE, CHAIN AND SYNTHETICS

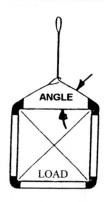

A CHOKER HITCH HAS 75%
OF THE CAPACITY OF A
SINGLE LEG ONLY IF THE
CORNERS ARE SOFTENED
AND THE HORIZONTAL
ANGLE IS GREATER THAN
30 DEGREES USE BLOCKS
TO PREVENT ANGLES LESS
THAN 30 DEGREES

BASKET HITCHES

WIRE ROPE

A BASKET HITCH HAS
TWICE THE CAPACITY
OF A SINGLE LEG ONLY
IF D/d RATIO IS 25/1 AND
IT IS VERTICAL.

WIRE ROPE, CHAIN AND SYNTHETICS

ANGLES DEGREES	PERCENTAGE OF SINGLE LEG CAPACITY
90	
60	
45	200%
30	170%
	140%
	100%

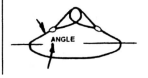

Table 3.9. Wire Rope Sling Capacities [1]

6 X 19 AND 6 X 37
IMPROVED PLOW STEEL – IWRC 5/1 DESIGN FACTOR

WIRE ROPE SIZE	Q & T CARBON SHACKLE MINIMUM SHACKLE SIZE FOR A D/d > 1 AT LOAD CONNECTION — SHACKLE SIZE	VERTICAL (SINGLE LEG)	CHOKER	TWO LEG OR BASKET HITCH 90°	60 DEGREE SLING ANGLE 60°	45 DEGREE SLING ANGLE 45°	30 DEGREE SLING ANGLE 30°
1/4	5/16	1120	820	2200	1940	1500	1100
5/16	3/8	1740	1280	3400	3000	2400	1700
3/8	7/16	2400	1840	4800	4200	3400	2400
7/16	1/2	3400	2400	6800	5800	4800	3400
1/2	5/8	4400	3200	8800	7600	6200	4400
9/16	5/8	5600	4000	11200	9600	7900	5600
5/8	3/4	6800	5000	13600	11800	9600	6800
3/4	7/8	9800	7200	19600	16900	13800	9800
7/8	1	13200	9600	26400	22800	18600	13200
1	1-1/8	17000	12600	34000	30000	24000	17000
1-1/8	1-1/4	20000	15800	40000	34600	28300	20000
1-1/4	1-3/8	26000	19400	52000	45000	36700	26000
1-3/8	1-1/2	30000	24000	60000	52000	42400	30000

- RATED CAPACITIES BASED ON PIN DIAMETER OR HOOK NO LARGER THAN THE NATURAL EYE WIDTH (1/2 X EYE LENGTH) OR LESS THAN THE NOMINAL SLING DIAMETER

REFER TO ANSI B30.9 FOR FULL DETAILS

HORIZONTAL SLING ANGLES OF LESS THAN 30 DEGREES ARE NOT RECOMMENDED

OPERATING PRACTICES / GUIDELINES FOR WIRE ROPE SLINGS

Operating practices and guidelines for the use of wire rope slings are as follows:

- Slings having suitable characteristics for the type of load, hitch, and environment shall be selected.
- The weight of the load shall be within the rated capacity of the sling.
- Wire rope slings shall not be shortened or lengthened by knotting or twisting or with wire rope clips or other methods not approved by the sling manufacturer.
- Slings that appear to be damaged shall not be used unless they are inspected and accepted as usable in accordance with the periodic inspection requirements stated above.
- The sling shall be hitched in a manner providing control of the load.
- Sharp corners in contact with the wire rope sling should be padded to minimize damage to the sling.
- Portions of the human body should be kept from between the sling and the load and from between the sling and the crane hook or hoist hook.
- Personnel should stand clear of the suspended load.
- Personnel shall not ride the sling.
- Shock loading is prohibited.
- Slings should not be pulled from under a load when the load is resting on the sling.
- Wire rope slings should be stored in an area where they will not be subjected to mechanical damage, corrosive action, moisture, extreme heat, or kinking.
- Twisting and kinking the legs shall be avoided.
- The load applied to the hook should be centered in the base (bowl) of the hook to prevent point loading of the hook, unless the hook is designed for point loading.
- During lifting, with or without load, personnel shall be alert for possible snagging.
- In a basket hitch, the load should be balanced to prevent slippage.
- The sling's legs should contain or support the load so that the load remains under control.
- Multiple-leg slings shall be selected so as not to introduce a working load in direct tension in any leg greater than that permitted. Triple- and quadruple-leg sling ratings should be considered the same as a double-sling rating because in normal lifting practice the load will not be uniformly distributed on all legs, leaving only two legs to carry the load. If rigging techniques—verified by a qualified rigger or rigging specialist—ensure the load is evenly distributed, then full use of three legs is allowed. Special rigging techniques verified by a qualified engineer shall be required to prove a load is evenly distributed over four or more sling legs.
- Slings should be long enough so that the rated load is adequate when the angle of the legs is taken into consideration.
- Slings should not be dragged on the floor or over an abrasive surface.

- In a choker hitch, slings shall be long enough so that the choker fitting chokes on the wire rope body and never on the sling fitting.

- Slings shall not be inspected by passing bare hands over the wire rope body. Broken wires, if present, may injure the hands.

- Fiber core wire rope should not be subjected to degreasing or a solvent because it will damage the core.

- Single-leg slings with hand-tucked splices can be unlaid by rotation. Care should be taken to minimize rotation.

- An object engaging the eye of a loop eye sling should not be greater in width than one-half the length of the loop eye.

Basic Inspection Criteria

The goal of a sling inspection is to evaluate remaining strength in a sling that has been used previously to determine if it is suitable for continued use. Specific inspection intervals and procedures are required by OSHA and by ASME B30.9 regulations, and the responsibility for performance of inspections is placed squarely upon the sling user by federal legislation.

As a starting point, the same work practices that apply to all "working" wire rope apply to wire rope that has been fabricated into a sling. Therefore, a good working knowledge of wire rope design and construction is not only useful but essential in conducting a wire rope sling inspection. But because wire rope is rather complex, no precise rules can be given to determine exactly when a wire rope sling should be replaced. There are many variables, and all must be considered.

OSHA standards specify that a wire rope sling shall be removed from service immediately if ANY of the following conditions are present:

- *Broken Wires*: For strand-laid grommets and single part slings, ten randomly distributed broken wires in one rope lay, or five broken wires in one strand in one rope lay. For cable laid, cable-laid grommets, and multi-part slings, use Table 3.10.

- *Metal Loss*: Wear or scraping of one-third the original diameter of outside individual wires. This is quite difficult to determine on slings and experience should be gained by the inspector by taking apart old slings and actually measuring wire diameters.

Table 3.10. Broken Wires

Sling body	Allowable broken wires	
	per lay	per braid
cable laid	20	
less than 8 part braid		20
8 part braid or more		40

- *Distortion*: Kinking, crushing, birdcaging, or other damage which distorts the rope structure. The main thing to look for is wire or strands that are pushed out of their original positions in the rope. Slight bends in a rope where ropes or strands are still relatively in their original positions would not be considered serious damage, but good judgment is indicated.

- *Heat Damage*: Any metallic discoloration or loss of internal lubricant caused by exposure to heat.

- *Bad End Attachments*: Cracked, bent, or broken end fittings caused by abuse, wear, or accident.

- *Bent Hooks*: No more than 15 percent over the normal throat openings is permissible, measured at the narrowest point, or twisting of more than 10 degrees.

- *Metal Corrosion*: Severe corrosion of the rope or end attachments that has caused pitting or binding of wires should be cause for replacing the sling. Light rusting usually does not affect strength of a sling, however.

- *Pulled Eye Splices*: Any evidence that eye spices have slipped, tucked strands have moved, or pressed sleeves show serious damage may be sufficient cause to reject a sling.

- *Unbalance*: A very common cause of damage is the kink that results from pulling through a loop while using a sling, thus causing wires and strands to be deformed and pushed out of their original position. This unbalances the sling, reducing its strength.

Disposition of Retired Slings

The best inspection program available is of no value if slings that are worn out and have been retired are not disposed of properly. When it is determined by the inspector that a sling is worn out or damaged beyond use, it should be immediately tagged "DO NOT USE." This sling should be destroyed as soon as possible by cutting the eyes and fitting from the rope with a torch. This will ensure that an employee will not mistakenly use a sling that has been retired from service.

Table 3.11. Wire Rope Sling Fact [1]

INSPECTION

ALL SLINGS SHALL BE VISUALLY INSPECTED BY THE PERSON HANDLING THE SLING EACH DAY THEY ARE USED: IN ADDITION, A PERIODIC INSPECTION SHALL BE PERFORMED BY A DESIGNATED PERSON, AT LEAST ANNUALLY, AND SHALL INCLUDE A RECORD OF THE INSPECTION.

- DISTORTION OF THE ROPE IN THE SLING SUCH AS KINKING, CRUSHING, UNSTRANDING, BIRDCAGING, MAIN STRAND DISPLACEMENT OR CORE PROTRUSION. LOSS OF ROPE DIAMETER IN SHORT ROPE LENGTHS OR UNEVENNESS OF OUTER STRANDS SHOULD PROVIDE EVIDENCE THE SLING SHOULD BE REPLACED.
- GENERAL CORROSION
- BROKEN OR CUT STRANDS
- NUMBER, DISTRIBUTION, AND TYPE OF VISIBLE BROKEN WIRES

REPLACEMENT

CONDITION SUCH AS THE FOLLOWING SHOULD BE SUFFICIENT REASON FOR CONSIDERATION OF SLING REPLACEMENT

- FOR STRAND LAID AND SINGLE PART SLINGS TEN RANDOMLY DISTRIBUTED BROKEN WIRES IN ONE ROPE LAY, OR FIVE BROKEN WIRES IN ONE ROPE STRAND IN ONE ROPE LAY.
- SEVERE LOCALIZED ABRASION OR SCRAPING
- KINKING, CRUSHING, BIRDCAGING, OR ANY DAMAGE RESULTING IN DISTORTION OF THE ROPE STRUCTURE.
- EVIDENCE OF HEAT DAMAGE
- END ATTACHMENTS THAT ARE CRACKED, DEFORMED, OR WORN TO THE EXTENT THAT THE STRENGTH OF THE SLING IS SUBSTANTIALLY AFFECTED.
- HOOKS SHOULD BE INSPECTED IN ACCORDANCE WITH ANSI B30.10
- SEVERE CORROSION OF THE ROPE OR END ATTACHMENTS

MULTI-PART REMOVAL CRITERIA FOR CABLE LAID AND BRAIDED SLINGS

SLING BODY	ALLOWABLE BROKEN WIRE PER LAY OR ONE BRAID	ALLOWABLE BROKEN STRANDS PER SLING LAY
LESS THAN 8 PER BRAID	20	1
CABLE LAID	20	1
8 PARTS AND MORE	40	1

OPERATING PRACTICES FOR STEEL CHAINS

Operating practices and criteria for the use of alloy steel chains are as follows:

- The weight of the load must be established and be within the rated load (working load limit) of the sling.

- Slings must be selected that match the load and environment in which they are to be used.

- The load should never be loaded on the point of the hook unless the hook is designed for such loading.

- The sling must be hitched or rigged in a manner that will control the load.

- Chain slings shall not be shortened or lengthened by knotting, twisting, or other methods not approved by the sling manufacturer.

- Slings that appear to be damaged shall not be used unless they are inspected and accepted as usable in accordance with the periodic inspection requirements stated above.

- Sharp corners in contact with the chain sling should be padded with material of sufficient strength to minimize damage to the sling.

- Portions of the human body should be kept from between the sling and the load and from between the sling and the crane/hoist hook.

- Personnel should stand clear of the suspended load.

- Personnel shall not ride the sling.

- Shock loading is prohibited.

- Slings should not be pulled from under a load when the load is resting on the sling.

- Slings should be stored in an area where they will not be subjected to mechanical damage, corrosive action, moisture, extreme heat, or kinking.

- Twisting and kinking the legs (branches) shall be avoided.

- During lifting, with or without load, personnel shall be alert for possible snagging.

- In basket hitch, the load should be balanced to prevent slippage.

- The sling's legs (branches) should contain or support the load so that the load remains under control.

- Multiple-leg (branch) chain slings shall be selected accordingly when used at specific angles. Operation at other angles shall be limited to rated loads of the next lower angle given in the appropriate table or calculated so as to not introduce into the leg (branch) itself a working load in direct tension greater than that permitted.

- Slings should be long enough so that the rated load is adequate when the angle of the legs (branches) is taken into consideration.

Table 3.12. Chain Sling Capacities [1]

CHAIN SIZE	VERTICAL (SINGLE LEG)	TWO LEG OR BASKET HITCH (90°)	60 DEGREE SLING ANGLE	45 DEGREE SLING ANGLE	30 DEGREE SLING ANGLE	Q T ALLOY MASTER LINK SIZE SINGLE LEG	Q T ALLOY MASTER LINK SIZE DOUBLE LEG
CHAIN GR – 8 DESIGN FACTOR 4/1							
1/4 - (9/32)	3500	7000	6050	4900	3500	1/2	1/2
3/8	7100	14200	12200	10000	7100	3/4	3/4
1/2	12000	24000	20750	16950	12000	1	1
5/8	18100	39200	31350	25500	18100	1	1-1/4
3/4	28300	56600	49000	40000	28300	1-1/4	1-1/2
7/8	34200	68400	59200	48350	34200	1-1/2	1-3/4
1	47700	95400	82600	67450	47700		
1-1/4	72300	144600	125200	102200	72300		

TABLE 1
MAXIMUM ALLOWABLE WEAR AT ANY POINT OF LINK

NORMAL CHAIN OR COUPLING LINK CROSS SECTION	MAXIMUM ALLOWABLE WEAR DIAMETER INCHES
9/32	.037
3/8	.052
1/2	.069
5/8	.084
3/4	.105
7/8	.116
1	.137
1-1/4	.169

REFER TO ANSI B30.9 FOR FULL DETAILS
HORIZONTAL SLING ANGLES OF LESS THAN 30 DEGREES ARE NOT RECOMMENDED

CHAIN – FACTS

INSPECTION AND REMOVAL FROM SERVICE PER ANSI N30.9

FREQUENT INSPECTION

• DAILY CHECK CHAIN AND ATTACHMENTS FOR WEAR, NICKS, CRACKS, BREAKS, GOUGES, STRETCH, BENDS, WELD SPLATTER, DISCOLORATION FROM EXCESSIVE TEMPERATURE, AND THROAT OPENINGS OF HOOKS.

1. CHAIN LINKS AND ATTACHMENTS SHOULD HINGE FREELY TO ADJACENT LINKS.
2. LATCHES ON HOOKS, IF PRESENT SHOULD HINGE FREELY AND SEAT PROPERLY WITHOUT EVIDENCE OF PERMANENT DISTORTION.

PERIODIC INSPECTION - INSPECTION RECORDS REQUIRED

• NORMAL SERVICE - YEARLY
• SEVERE SERVICE - MONTHLY

THIS INSPECTION SHALL INCLUDE EVERYTHING IN A FREQUENT INSPECTION PLUS EACH LINK AND END ATTACHMENT SHALL BE EXAMINED INDIVIDUALLY, TAKING CARE TO EXPOSE INNER LINK SURFACES OF THE CHAIN AND CHAIN ATTACHMENTS

1. WORN LINKS SHOULD NOT EXCEED VALUES GIVEN IN TABLE 1 OR RECOMMENDED BY THE MANUFACTURER
2. SHARP TRANSVERSE NICKS AND GOUGES SHOULD BE ROUNDED OUT BY GRINDING AND THE DEPTH OF THE GRINDING SHOULD NOT EXCEED VALUES IN TABLE 1
3. HOOKS SHOULD BE INSPECTED IN ACCORDANCE WITH ANSI B30.10
4. IF PRESENT, LATCHES ON HOOKS SHOULD SEAT PROPERLY, ROTATE FREELY, AND SHOW NO PERMANENT DISTORTION

- Slings should not be dragged on the floor or over an abrasive surface.

- When used in a choker hitch arrangement, slings shall be selected to prevent the load on any portion of the sling from exceeding the rated load of the chain sling components.

- Before using a chain sling outside the temperature range of -40° F to 400° F (-20° C to 204° C), contact the sling manufacturer.

HOOKS, SHACKLES, AND EYEBOLTS

Hooks

Wire rope, shackles, rings, master links, and other rigging hardware must be capable of supporting, without failure, at least five times the maximum intended load applied or transmitted to that component. The theoretical reserve capacity of a hoist hook should be a minimum of 5 to 1 for carbon eye hooks, alloy eye hooks, and carbon shank hooks, and 4.5 to 1 for an alloy shank hook. Known as the design factor, it is usually computed by dividing the given ultimate load by the working load limit. The ultimate load is the average load or force at which the product fails or no longer supports the load. The working load is the maximum mass or force that the product is authorized to support in general service. The design factor is generally expressed as a ratio such as 5 to 1. Also important to the design of hooks is the selection of proper steel.

Quenching and tempering ensures the uniformity of performance and maximizes the properties of the steel. This means that each hook meets its rated strength and other properties. This quenching and tempering process develops a tough material that reduces the risk of a brittle, catastrophic failure, thus improving impact and fatigue properties. As a result, if overloaded, the hook will deform before ultimate failure occurs, thus giving warning. The requirements of hoisting and rigging jobs associated with cranes demand this reliability and consistency. Quench and tempering ensures that not only is the working load limit met, but that ductility, fatigue, and impact properties are appropriate. Welding is not permitted on hooks in any field application.

The proper application of hoist hooks requires that the correct type, size, and working load capacity of hooks be used. All hooks must be load rated (with either the working load or a cross reference code). In addition the traceability code, size, and manufacturer's name should be boldly marked on the product. Availability of a full line of eye, shank, and swivel hooks in carbon and steel is essential when selecting the desired hook for the proper application. Detailed application information is needed in the proper selection and use of hooks. This type information is most effective when it is provided in supporting brochures and engineering information. A formal application and warning system is needed to attract the attention of the user, clearly informing the user of the factors involved in the task, and the proper application procedures.

All new and repaired hooks require an inspection to verify compliance with ANSI B30.10, Hooks prior to initial use. A record of the initial inspection must be made. Once hooks are placed in service, there are two general inspection classifications based upon intervals at which examination shall be performed: The classifications are designated frequent and periodic, with intervals between examinations as defined below.

Frequent Inspection Classification Intervals

Visual examinations by the operator or other employer-designated person must occur at the following intervals:

- Normal service: monthly
- Heavy service: weekly to monthly
- Severe service: daily to weekly.

No records are required for the visual examinations.

Hook Service Classifications

The service classifications identified in the above frequent inspection classification intervals are as follows:

- Normal service: Service that involves operating at less than 85% of rated load except for isolated instances
- Heavy service: Service that involves operating at 85% to 100% of rated load as a regular specified procedure
- Severe service: Heavy service coupled with abnormal operating conditions.

Frequent inspections include:

- Observations during operation of the hook
- Determination whether any conditions constitute a hazard and whether a more detailed inspection is required.
- The following items are to be inspected:
  - Distortion, such as bending, twisting, or increased throat opening
  - Wear
  - Cracks, nicks, or gouges
  - Latch engagement, damaged or malfunctioning latch (if provided)
  - Hook attachment and securing means.

Periodic Inspection Classification Intervals

For periodic inspections, visual inspections are to be performed by a qualified inspector for the following three intervals. Any apparent external conditions detected are to be recorded to provide the basis for continuing evaluation:

- Normal service: yearly, with equipment in place

- Heavy service: semiannually, with equipment in place unless external conditions indicate that disassembly should be done to permit detailed inspection

- Severe service: quarterly, as in heavy service, except that detailed inspection may show the need for nondestructive type testing.

Periodic Inspection

Periodic inspections include the requirements of frequent inspection listed above. Hooks having any of the following conditions are required to be removed from service until repaired or replaced:

- Deformation: Any bending or twisting exceeding 10 degrees from the plane of the unbent hook

- Throat Opening: Any distortion causing an increase in throat opening exceeding 15 percent

- Wear: Any wear exceeding 10 percent of the original section dimension of the hook or its load pin.

Inspection Records

Following is a summary of inspection record requirements.

1. Initial Inspection: A record of the initial inspection shall be made

2. Pre-use and Frequent Inspection: No records are required

3. Periodic Inspection: A record of periodic inspections shall be made.

Inspection records should include the:

- Date of inspection

- The signature of the person who performed the inspection

- The serial number or other identifier of the hook inspected.

Inspection records are to be retained as part of the equipment history record.

Shackles

The theoretical reserve capacity of carbon shackles should be a minimum 5 to 1, and alloy shackles a minimum of 5 to 1. Known as the *design factor*, it is usually computed by dividing the provided ultimate load by the working load limit. The ultimate load is the average load or force at which the product fails or no longer supports the load. The working load limit is the maximum mass or force that the product is authorized to support in general service. The design factor is generally expressed as a ratio such as 5 to 1. Also important to the design of shackles is the selection of proper steel to support fatigue, ductility, and impact properties.

The proper performance of premium shackles depends on good manufacturing techniques that include proper forging and accurate machining. Closed die forging of shackles ensures clear lettering, superior grain flow, and consistent dimensional accuracy. A closed die forged bow allows for an increased cross section that, when coupled with quench and tempering, enhances strength and durability. Closed die bow forgings combined with close tolerance pinholes ensure good fatigue life. Close pin-to-hole tolerance has been proven to be critical for good fatigue life, particularly with screw pin shackles.

As with hooks, quench and tempering ensures that each shackle meets its rated strength and all other essential properties. These properties ensure the shackle bow will deform if overloading occurs, giving warning before ultimate failure. The proper application of shackles requires that the correct type and size of shackles be used. The shackle's working load limit, its size, a traceability code, and the manufacturer's name should be clearly and boldly marked in the bow. Traceability of the material chemistry and properties is essential to total confidence in the product.

Inspection Criteria

Before each use, shackles are required to be inspected to ensure that:

- Shackle pins fit freely without binding (seated screw pin shackles shall be disassembled by hand after the first-half turn.) The shackle must have no defect that will interfere with serviceability and the pin must not show signs of deformation.

- Shackle pins must fit freely without binding. (Seated screw pin shackles can be disassembled by hand after the first-half turn.)

- There is no damage, corrosion, wear, cracks, or twists.

Table 3.13. Rigging Hardware - Shackles/Hooks [1]

SHACKLES

SCREW PIN AND BOLT TYPE — CARBON SHACKLE DESIGN FACTOR 6/1 — ALLOY SHACKLE DESIGN FACTOR 5/1 — QUENCHED AND TEMPERED QUIC-CHECK

NOMINAL SIZE (IN) DIAMETER OF BOWS	CARBON MAXIMUM WORKING LOAD TONS	ALLOY MAXIMUM WORKING LOAD TONS	INSIDE WIDTH AT PIN (INCHES)	DIAMETER OF PIN
3/16	1/3		.38	.25
1/4	1/2		.47	.31
5/16	3/4		.53	.38
3/8	1	2	.66	.44
7/16	1-1/2	2.6	.75	.50
1/2	2	3.3	.81	.63
5/8	3-1/4	5	1.06	.75
3/4	4-3/4	7	1.25	.88
7/8	6-1/2	9.5	1.44	1.00
1	8-1/2	12.5	1.69	1.13
1-1/8	9-1/2	15	1.81	1.25
1-1/4	12	18	2.03	1.38
1-3/8	13-1/2	21	2.25	1.50
1-1/2	17	30	2.38	1.63

- INSURE SCREW PIN TIGHT BEFORE EACH LIFT
- USE BOLT TYPE SHACKLE FOR PERMANENT INSTALLATION
- DO NOT SIDE LOAD ROUND PIN SHACKLE
- USE SCREW PIN OR BOLT TYPE TO COLLECT SLINGS

MAXIMUM INCLUDED ANGLE 120 DEGREES

HOOKS

SHANK HOOK / SWIVEL HOOK / EYE HOOK

DESIGN FACTOR
- EYEHOOKS – 5/1 (EXCEPT ALLOY 30 TON AND LARGER ARE 4-1/2 /1
- SHANK AND SWIVEL ARE 4-1/2 /1

QUENCHED AND TEMPERED QUIC-CHECK

CARBON MAXIMUM WORKING LOAD TONS	CODE	ALLOY MAXIMUM WORKING LOAD TONS	CODE	THROAT OPENING (INCHES)	DEFORMATION INDICATOR A - A
3/4	DC	1	DA	.88	1.50
1	FC	1-1/2	FA	.97	2.00
1-1/2	GC	2	GA	1.00	2.00
2	HC	3	HA	1.12	2.00
3	IC	*4-1/2 /5	IA	1.06	2.50
5	JC	7	JA	1.50	3.00
7-1/2	KC	11	KA	1.75	4.00
10	LC	15	LA	1.91	4.00
15	NC	22	NA	2.75	5.00
20	OC	30	OA	3.25	6.50
25	PC	37	PA	3.00	7.00
30	SC	45	SA	3.38	8.00
40	TC	60	TA	4.12	10.00

* 320N EYE HOOK IS NOW RATED AT 5 TONS

THROAT OPENING

MAXIMUM INCLUDED ANGLE 90 DEGREES

EYE HOOK

- DO NOT SIDELOAD
- DO NOT TIP LOAD
- DO NOT BACKLOAD

Eyebolts

New equipment may be delivered with eyebolts installed that do not meet the requirements for lifting the equipment. To ensure that such eyebolts can be used, a determination must be made if the installed eyebolts are acceptable for their intended use. The vertical safe-working load should be identified on the eyebolts by the manufacturer.

Before using any eyebolt, a visual inspection in addition to the safe-working load must be made to ensure that there has been no deformation. Additionally, a determination of the condition of the threads and shank should be made before using. The threads on the shank and the receiving hole must be clean. Cutting, grinding, or machining an eyebolt is not permitted. The eyebolt must be screwed down completely and the nuts tightened securely against the load for proper seating.

For angular lifts, shoulder nut or machinery eyebolts must be used. The working load on angular lifts is adjusted according to the angle of the pull. If the direction of pull is 45 degrees, the adjusted working load would be 30% of the rated working load. For a 90 degree pull, the adjusted working load would be 25% of the rated working load. Table 3.14 provides the safe working loads of eyebolts corresponding to the angle of the pull.

Table 3.14. Safe Loading of Shoulder-Type Eyebolts*

SAFE WORKING LOADS CORRESPONDING TO ANGLE OF PULL

Diameter (in.)	Vertical	75 degrees	60 degrees	45 degrees	Less than 45 degrees
1/4	500				
5/16	800	Reduce	Reduce	Reduce	
3/8	1,200	vertical	vertical	vertical	
1/2	2,200	loads	loads	loads	NOT RECOMMENDED
5/8	3,500	by	by	by	
3/4	5,200	45%	65%	75%	
7/8	7,200				
1	10,000				
1-1/4	15,200				
1-1/2	21,400				

* The safe working load for eyebolts with no shoulders is the same as above under a vertical load.

Table 3.15. Rigging Hardware - Shoulder Eyebolts/Swivel Hoist Rings [1]

WIRE ROPE CLIPS

SIZE	EFFICIENCY	NUMBER OF CLIPS	TURNBACK LENGTH (IN)	TORQUE FT – lbs
1/8	80%	2	3-1/4	4.5
3/16	80%	2	3-3/4	7.5
1/4	80%	2	4-3/4	15
5/16	80%	2	5-1/4	30
3/8	80%	2	6-1/2	45
7/16	80%	2	7	65
1/2	80%	3	11-1/2	65
9/16	80%	3	12	95
5/8	80%	3	12	95
3/4	80%	4	18	130
1	90%	5	26	225

APPLY U-BOLT OVER DEAD END OF THE WIRE ROPE
LIVE END OF THE ROPE RESTS IN THE SADDLE
A TERMINATION IS NOT COMPLETE UNTIL IT HAS BEEN
RETORQUED A SECOND TIME
NEVER SADDLE A DEAD HORSE!

TURNBUCKLE

SIZE	WORKING LOAD LIMIT JAW AND EYE 5/1 DESIGN FACTOR	WORKING LOAD LIMIT HOOK END FITTING 5/1 DESIGN FACTOR
1/4	500	400
5/16	800	700
3/8	1200	1000
1/2	2200	1500
5/8	3500	2250
3/4	5200	3000
7/8	7200	4000
1	10000	5000
1-1/4	15200	5000
1-1/2	21400	7500

THE USE OF LOCKNUTS OR MOUSING IS AN EFFECTIVE METHOD OF PREVENTING TURNBUCKLES FROM ROTATING

NATURAL AND SYNTHETIC FIBER ROPE SLINGS

Usage of Fiber Ropes

Unreeling

Remove rope properly from reels to prevent kinking. The rope should be removed by pulling it off the top while the reel is free to rotate. To proceed in any other manner may cause kinks or strand distortion.

Handling

Never stand in line with rope under tension. If a rope fails, it can recoil with lethal force. Synthetic rope has higher recoil tendencies than natural fiber rope. Reverse rope ends regularly. This permits even wearing and ensures a longer, useful life. When using tackle or slings, apply a steady, even pull to get full strength from rope.

Overloading

Because of the wide range of rope use, exposure to several factors affecting rope behavior, and the degree of risk to life and property involved, it is impossible to make blanket recommendations as to working loads. However, to provide guidelines, working loads are tabulated for rope in good condition with appropriate splices, in non-critical applications, and under normal service conditions.

A higher working load may be selected only with expert knowledge of conditions and professional estimate of risk and if the rope meets the following criteria:

- Has not been subject to dynamic loading or other excessive use

- Has been inspected and found to be in good condition

- Is to be used in the recommended manner

- The application does not involve elevated temperatures, extended periods under load, or obvious dynamic loading (see below) such as sudden drops, snubs, or pickups.

For all such applications and for applications involving more severe exposure conditions, or for recommendations on special applications, consult the manufacturer. Many uses of rope involve serious risk of injury to personnel or damage to property. This danger is often obvious, as when a heavy load is supported close to one or more workers. An equally dangerous situation occurs if personnel are in line with a rope under tension. Should the rope fail, it could recoil with lethal force. Employees must be warned against the serious danger of standing in line with any rope under tension.

In all cases where such risks are present, or there is any question about the loads involved or the conditions of use, the working load should be substantially reduced and the rope properly inspected.

Dynamic Loading

Normal working loads are not applicable when rope is subject to significant dynamic loading. Instantaneous changes in load, up or down, in excess of 10% of the line's rated working load constitute hazardous shock load and would void normal working loads. Whenever a load is picked up, stopped, or swung there is an increased force due to dynamic loading. The more rapidly or suddenly such actions occur, the greater the increase will be. In extreme cases, the force put on the rope may be two, three, or more times the normal load involved. Examples could be picking up a tow on a slack line or using a rope to stop a falling object. Therefore, in all such applications such as towing lines, lifelines, safety lines, climbing ropes, etc., working loads as given DO NOT APPLY.

Users should be aware that dynamic effects are greater on a low elongation rope such as polyester than on a high elongation rope such as nylon, and greater on a shorter rope than a longer one. Any working load ratios listed contain provision for very modest dynamic loads. This means, however, that when this working load has been used to select a rope, the load must be handled slowly and smoothly to minimize effects and avoid exceeding provision for them.

Abrasion

Avoid all abrasive conditions. All rope will be severely damaged if subjected to rough surfaces or sharp edges. Chocks, bits, winches, drums, and other surfaces must be kept in good condition and free of burrs and rust. Pulleys must be free to rotate and should be of proper size to avoid excessive wear. Clamps and similar devices will damage and weaken the rope and should be used with extreme caution. Do not drag rope over rough ground. Dirt and grit picked up by rope can work into the strands, cutting the inside fibers.

Chemicals

Avoid chemical exposure. Rope is subject to damage by chemicals. Consult the manufacturer for specific chemical exposure issues, such as solvents, acids, and alkalies. This is particularly true for natural fiber rope. Consult the manufacturer for recommendations when a rope will be used where chemical exposure (either fumes or actual contact) can occur.

Temperature

Effect on tensile strength. Tensile strength charts apply to ropes tested at normal room temperatures, 70° F. Ropes have lower tensile strengths at higher temperatures, 30% (or more) lower at the boiling point of water (212° F), and continuing on down to zero strength for nylon and polyester at 480° F, and 300° F for polypropylene. Also, continued exposure at elevated temperatures can melt and part synthetic ropes or cause permanent damage.

Splicing

Join rope by splicing. Knots can decrease rope strength by as much as 60%. Use the manufacturers recommended splices for maximum efficiency. Other terminations can be used, but their strength loss with particular types of rope and construction should be determined and not assumed.

Storage and Care

All rope should be stored clean, dry, out of direct sunlight, and away from extreme heat. Some synthetic rope (particularly polypropylene and polyethylene) may be severely weakened by prolonged exposure to ultraviolet (UV) rays unless specifically stabilized and/or pigmented to increase its UV resistance. UV degradation is indicated by discoloration and the presence of splinters and slivers on the surface of the rope.

Formulas to determine reel and storage capacities:

Use inch reel dimensions.

$$\text{Rope Length Feet} = \frac{(\text{Traverse Width})(\text{Flange Diameter}^2 - \text{Barrel Diameter}^2)}{(16)(\text{Rope Diameter}^2)}$$

Formulas to determine bin capacity: $V = C^2 \times L \times R$

Where V = volume in cubic inches

C = rope circumference in inches

L = length of rope in feet

R = 1.58 for carefully stored rope or 2.0 for random packing.

Inspection

Avoid using rope that shows signs of aging and wear. If there is any question or doubt, destroy the used rope. No type of visual inspection can be guaranteed to accurately and precisely determine actual residual strength. When the fibers show wear in any given area the rope should be respliced to eliminate the damaged area, be downgraded, or be replaced.

Check the line regularly for frayed strands and broken yarns. Pulled strands should be rethreaded into the rope if possible. A pulled strand can snag during a rope operation. Both outer and inner rope fibers contribute to the strength of the rope. When either is worn, the rope is weakened. A heavily used rope will often become compacted or hard, which indicates reduced strength. The dielectric strength of rope in this condition is also reduced.

Web Slings

Synthetic web slings are fabricated by sewing woven synthetic webbing of nylon or polyester yarns and have the following characteristics:

- Uniform thickness and width

- Sufficient certified tensile strength to meet the sling manufacturer's requirements

- Full woven width, including selvage edges

- Webbing ends sealed by heat, or other suitable means, to prevent raveling

- Stitching shall be the only method used to attach end fittings to webbing and to form eyes.

Synthetic web slings are required to be labeled (a sewn-on leather tag is recommended) with the label stating the following:

- Manufacturer's name or trademark

- Manufacturer's code or stock number

- Rated loads for the types of hitches used

- Type of synthetic web material

- It is up to the user to add an additional tag, sticker, or other identifier to indicate when the next periodic inspection is required.

A synthetic web sling cannot be used at a load greater than shown on its tag. Figure 3.6 provides the different sling capacities and the inspection and removal criteria per ANSI B30.9.

Figure 3.6. Web Sling Capacities [1]

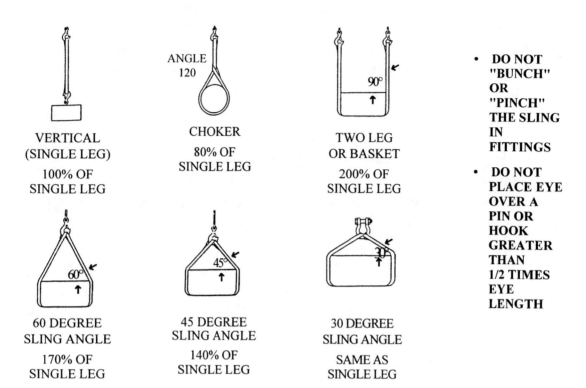

VERTICAL
(SINGLE LEG)

100% OF
SINGLE LEG

CHOKER
80% OF
SINGLE LEG

TWO LEG
OR BASKET

200% OF
SINGLE LEG

- DO NOT "BUNCH" OR "PINCH" THE SLING IN FITTINGS

- DO NOT PLACE EYE OVER A PIN OR HOOK GREATER THAN 1/2 TIMES EYE LENGTH

60 DEGREE
SLING ANGLE

170% OF
SINGLE LEG

45 DEGREE
SLING ANGLE

140% OF
SINGLE LEG

30 DEGREE
SLING ANGLE

SAME AS
SINGLE LEG

WEB SLING
INSPECTION AND REMOVAL FROM SERVICE PER ANSI B30.9

FREQUENT INSPECTION
> THIS INSPECTION SHALL BE MADE BY THE PERSON HANDLING THE SLING EACH DAY THE SLING IS USED

PERIODIC INSPECTION WRITTEN INSPECTION RECORDS SHOULD BE KEPT FOR ALL SLINGS
> THIS INSPECTION SHOULD BE CONDUCTED BY DESIGNATED PERSONNEL, FREQUENCY OF THE INSPECTION SHOULD BE BASED THE FOLLOWING:

1. FREQUENCY OF SLING USE
2. SEVERITY OF SERVICE CONDITIONS
3. EXPERIENCE GAINED ON THE SERVICE LIFE OF SLING USED IN SIMILAR APPLICATIONS
4. AT LEAST ANNUALLY

REMOVAL CRITERIA
1. ACID OR CAUSTIC BURNS
2. MELTING OR CHARRING OF ANY PART OF THE SLING
3. BROKEN, TEARS, CUTS, OR SNAGS
4. BROKEN OR WORN STITCHING IN LOAD BEARING SPLICES
5. EXCESSIVE ABRASIVE WEAR
6. KNOTS IN ANY PART OF THE SLING
7. EXCESSIVE PITTING OR CORROSION, OR CRACKED DISTORTED OR BROKEN FITTINGS
8. OTHER VISIBLE DAMAGE THAT CAUSES DOUBT AS TO THE STRENGTH OF THE SLING

REFER TO ANSI B30.9 FOR FULL DETAILS

HORIZONTAL SLING ANGLES OF LESS THAN 30 DEGREES ARE NOT RECOMMENDED

SUMMARY

All hoisting and rigging operations consist of three components: the crane that is used, the crane operator, and the rigging crew that is responsible for rigging the load and hooking it to the crane. All components must work in unison in order to accomplish the intended task. By working closely with the rigging crew, the operator can be assured that all rigging equipment has been inspected, correct slings and hardware have been selected, and the load is properly attached. To support the crew in making the correct choices when preparing to make a lift, there are several beneficial aids available from different companies.

Before any lift is made, the operator must know or be familiar with the following:

- The weight of the load to be lifted, or use a crane equipped with a load-indicating device on which he or she has been trained to use.

- The center of gravity: This is the location point of the load where the weight of the load is evenly distributed at all attachment points.

- Sling design and construction: To select the best wire rope or web sling for each application, one must know the required performance characteristics for the job and enough about the sling design to select the optimum combination of the sling properties.

REFERENCES

1. The Crosby Group, Inc. *Crosby User's Lifting Guide*, Tulsa, OK. 1998, pp. 1-12.

2. Wire Rope Technical Board. *Wire Rope Sling Users Manual*, Exeter, PA. 1997, pp. 5, 7, 51-58.

4

CRANES AND DERRICKS RELATED OSHA STANDARDS

This chapter contains the following six OSHA standards with the author's comments (italicized text) provided:

4.A. - 29 CFR 1910.179 - Overhead and Gantry Cranes [1]

4.B. - 29 CFR 1910.180 - Crawler, Locomotive, and Truck Cranes [2]

4.C. - 29 CFR 1910.181 - Derricks [3]

4.D. - 29 CFR 1926.550 - Cranes and Derricks [4]

4.E. - 29 CFR 1926.251 - Rigging Equipment for Material Handling [5]

4.F. - 29 CFR 1910.184 - Slings [6]

4.A. OSHA 1910.179 - Overhead and Gantry Cranes

(a) Definitions Applicable to This Section.

(a)(1)

A "crane" is a machine for lifting and lowering a load and moving it horizontally, with the hoisting mechanism an integral part of the machine. Cranes whether fixed or mobile are driven manually or by power.

(a)(2)

An "automatic crane" is a crane which when activated operates through a preset cycle or cycles.

(a)(3)

A "cab-operated crane" is a crane controlled by an operator in a cab located on the bridge or trolley.

(a)(4)

"Cantilever gantry crane" means a gantry or semigantry crane in which the bridge girders or trusses extend transversely beyond the crane runway on one or both sides.

(a)(5)

"Floor-operated crane" means a crane which is pendant or nonconductive rope controlled by an operator on the floor or an independent platform.

(a)(6)

"Gantry crane" means a crane similar to an overhead crane except that the bridge for carrying the trolley or trolleys is rigidly supported on two or more legs running on fixed rails or other runway.

(a)(7)

"Hot metal handling crane" means an overhead crane used for transporting or pouring molten material.

(a)(8)

"Overhead crane" means a crane with a movable bridge carrying a movable or fixed hoisting mechanism and traveling on an overhead fixed runway structure.

(a)(9)

"Power-operated crane" means a crane whose mechanism is driven by electric, air, hydraulic, or internal combustion means.

(a)(10)

A "pulpit-operated crane" is a crane operated from a fixed operator station not attached to the crane.

(a)(11)

A "remote-operated crane" is a crane controlled by an operator not in a pulpit or in the cab attached to the crane, by any method other than pendant or rope control.

(a)(12)

A "semigantry crane" is a gantry crane with one end of the bridge rigidly supported on one or more legs that run on a fixed rail or runway, the other end of the bridge being supported by a truck running on an elevated rail or runway.

(a)(13)

"Storage bridge crane" means a gantry type crane of long span usually used for bulk storage of material; the bridge girders or trusses are rigidly or non-rigidly supported on one or more legs. It may have one or more fixed or hinged cantilever ends.

(a)(14)

"Wall crane" means a crane having a jib with or without trolley and supported from a side wall or line of columns of a building. It is a traveling type and operates on a runway attached to the side wall or columns.

(a)(15)

"Appointed" means assigned specific responsibilities by the employer or the employer's representative.

(a)(16)

"ANSI" means the American National Standards Institute.

(a)(17)

An "auxiliary hoist" is a supplemental hoisting unit of lighter capacity and usually higher speed than provided for the main hoist.

(a)(18)

A "brake" is a device used for retarding or stopping motion by friction or power means.

(a)(19)

A "drag brake" is a brake which provides retarding force without external control.

(a)(20)

A "holding brake" is a brake that automatically prevents motion when power is off.

(a)(21)

"Bridge" means that part of a crane consisting of girders, trucks, end ties, footwalks, and drive mechanism which carries the trolley or trolleys.

(a)(22)

"Bridge travel" means the crane movement in a direction parallel to the crane runway.

(a)(23)

A "bumper" [buffer] is an energy absorbing device for reducing impact when a moving crane or trolley reaches the end of its permitted travel; or when two moving cranes or trolleys come in contact.

(a)(24)

The "cab" is the operator's compartment on a crane.

(a)(25)

"Clearance" means the distance from any part of the crane to a point of the nearest obstruction.

(a)(26)

"Collectors current" are contacting devices for collecting current from runway or bridge conductors.

(a)(27)

"Conductors, bridge" are the electrical conductors located along the bridge structure of a crane to provide power to the trolley.

(a)(28)

"Conductors, runway" [main] are the electrical conductors located along a crane runway to provide power to the crane.

(a)(29)

The "control braking means" is a method of controlling crane motor speed when in an overhauling condition.

(a)(30)

"Countertorque" means a method of control by which the power to the motor is reversed to develop torque in the opposite direction.

(a)(31)

"Dynamic" means a method of controlling crane motor speeds when in the overhauling condition to provide a retarding force.

(a)(32)

"Regenerative" means a form of dynamic braking in which the electrical energy generated is fed back into the power system.

(a)(33)

"Mechanical" means a method of control by friction.

(a)(34)

"Controller, spring return" means a controller which when released will return automatically to a neutral position.

(a)(35)

"Designated" means selected or assigned by the employer or the employer's representative as being qualified to perform specific duties.

(a)(36)

A "drift point" means a point on a travel motion controller which releases the brake while the motor is not energized. This allows for coasting before the brake is set.

(a)(37)

The "drum" is the cylindrical member around which the ropes are wound for raising or lowering the load.

(a)(38)

An "equalizer" is a device which compensates for unequal length or stretch of a rope.

(a)(39)

"Exposed" means capable of being contacted inadvertently. Applied to hazardous objects not adequately guarded or isolated.

(a)(40)

"Fail-safe" means a provision designed to automatically stop or safely control any motion in which a malfunction occurs.

(a)(41)

"Footwalk" means the walkway with handrail, attached to the bridge or trolley for access purposes.

(a)(42)

A "hoist" is an apparatus which may be a part of a crane, exerting a force for lifting or lowering.

(a)(43)

"Hoist chain" means the load bearing chain in a hoist.

> Note: Chain properties do not conform to those shown in ANSI B30.9-1971, Safety Code for Slings.

(a)(44)

"Hoist motion" means that motion of a crane which raises and lowers a load.

(a)(45)

"Load" means the total superimposed weight on the load block or hook.

(a)(46)

The "load block" is the assembly of hook or shackle, swivel, bearing, sheaves, pins, and frame suspended by the hoisting rope.

(a)(47)

"Magnet" means an electromagnetic device carried on a crane hook to pick up loads magnetically.

(a)(48)

"Main hoist" means the hoist mechanism provided for lifting the maximum rated load.

(a)(49)

A "man trolley" is a trolley having an operator's cab attached thereto.

(a)(50)

"Rated load" means the maximum load for which a crane or individual hoist is designed and built by the manufacturer and shown on the equipment nameplate(s).

(a)(51)

"Rope" refers to wire rope, unless otherwise specified.

(a)(52)

"Running sheave" means a sheave which rotates as the load block is raised or lowered.

(a)(53)

"Runway" means an assembly of rails, beams, girders, brackets, and framework on which the crane or trolley travels.

(a)(54)

"Side pull" means that portion of the hoist pull acting horizontally when the hoist lines are not operated vertically.

(a)(55)

"Span" means the horizontal distance center to center of runway rails.

(a)(56)

"Standby crane" means a crane which is not in regular service but which is used occasionally or intermittently as required.

(a)(57)

A "stop" is a device to limit travel of a trolley or crane bridge. This device normally is attached to a fixed structure and normally does not have energy absorbing ability.

(a)(58)

A "switch" is a device for making, breaking, or for changing the connections in an electric circuit.

(a)(59)

An "emergency stop switch" is a manually or automatically operated electric switch to cut off electric power independently of the regular operating controls.

(a)(60)

A "limit switch" is a switch which is operated by some part or motion of a power-driven machine or equipment to alter the electric circuit associated with the machine or equipment.

(a)(61)

A "main switch" is a switch controlling the entire power supply to the crane.

(a)(62)

A "master switch" is a switch which dominates the operation of contactors, relays, or other remotely operated devices.

(a)(63)

The "trolley" is the unit which travels on the bridge rails and carries the hoisting mechanism.

(a)(64)

"Trolley travel" means the trolley movement at right angles to the crane runway.

(a)(65)

"Truck" means the unit consisting of a frame, wheels, bearings, and axles which supports the bridge girders or trolleys.

(b) General Requirements

(b)(1)

Application. This section applies to overhead and gantry cranes, including semigantry, cantilever gantry, wall cranes, storage bridge cranes, and others having the same fundamental characteristics. These cranes are grouped because they all have trolleys and similar travel characteristics.

(b)(2)

New and existing equipment. All new overhead and gantry cranes constructed and installed on or after August 31, 1971, shall meet the design specifications of the American National Standard Safety Code for Overhead and Gantry Cranes, ANSI B30.2.0-1967, which is incorporated by reference as specified in Sec. 1910.6.

> *The question is often asked if the preceding paragraph, (b)(2), implies that only the design specifications of ANSI B30.2.0 must be met and that only the 1967 revision applies and not any subsequent revisions. The answer is that only the design specifications of ANSI B30.2.0 apply to all new overhead and gantry cranes constructed and installed on or after August 31, 1971. OSHA accepts employers' use of the current revision to national consensus standards, such as ANSI B30.2.0, in place of earlier revisions incorporated by reference or adopted into OSHA standards. This acceptance is predicated on using the current revision, such that at least the same level of safety and health is provided, as required by complying with OSHA standards.*

(b)(3)

Modifications. Cranes may be modified and re-rated provided such modifications and the supporting structure are checked thoroughly for the new rated load by a qualified engineer or the equipment manufacturer. The crane shall be tested in accordance with paragraph (k) (2) of this section. New rated load shall be displayed in accordance with subparagraph (5) of this paragraph.

> *Section 1910.179(b)(3) relates to modification, after which the crane shall be tested in accordance with requirement of paragraph (k)(2). Cranes may be modified and re-rated provided such modifications and the supporting structure are checked thoroughly for the new rated load by a qualified engineer or the equipment manufacturer. The crane shall be tested in accordance with paragraph (k)(2) of this section. New rated load shall be displayed in accordance with subparagraph (5) of this paragraph. Section (b)(5) of this standard requires that each crane shall be plainly marked on each side with its rated load.*

> *As mentioned above, existing cranes constructed and installed prior to August 31, 1971, are not required to meet the specification of (b)(2). However, cranes constructed prior to this date and later modernized are also exempted from the specification of (b)(2), but such cranes must meet the requirements of this section (b)(3), Modifications.*

(b)(4)

Wind indicators and rail clamps. Outdoor storage bridges shall be provided with automatic rail clamps. A wind-indicating device shall be provided which will give a visible or audible alarm to the bridge operator at a predetermined wind velocity. If the clamps act on the rail heads, any beads or weld flash on the rail heads shall be ground off.

Section 1910.179(b)(4) requires that outdoor storage bridges shall be provided with automatic rail clamps and a wind-indicating device shall be provided which will give a visible or audible alarm to the bridge operator at a predetermined wind velocity. Wind indicators are only required on outdoor "storage bridges." The standard is concerned with the action that may be created by a wind force above that which it was designed to operate. The wind indicators are not required on an overhead traveling crane—only on "Gantry" type.

(b)(5)

Rated load marking. The rated load of the crane shall be plainly marked on each side of the crane, and if the crane has more than one hoisting unit, each hoist shall have its rated load marked on it or its load block and this marking shall be clearly legible from the ground or floor.

The statement in this paragraph, 1910.179(b)(5), concerning rated load markings on overhead bridge and gantry cranes, often causes questions as to whether capacity markings would be required on the hoist, trolley, and bridge individually, and, if so, would the markings be required to have the same load rating on each component.

If the crane has one hoisting unit, compliance with this standard is met when the rated load is plainly marked on each side of the crane at a location such as the bridge. In addition to marking the rated load for the bridge or gantry crane, if the crane has more than one hoisting unit, the employer can choose to mark the applicable rated load on the hoist or the load block. The rated load marking on a hoist must be located and arranged so that it is evident to the personnel responsible for the safe operation of the hoisting unit. Hoisting units on overhead and gantry cranes may have different rated loads. However, the crane must not be loaded beyond its rated load except for test purposes as provided in 1910.179(k). Section 1910.179(k) relates to testing of overhead cranes prior to initial use or whenever altered, and when testing to determine and confirm the maximum rated load the crane shall be permitted to handle.

(c) Cabs

(c)(1)(i)

The general arrangement of the cab and the location of control and protective equipment shall be such that all operating handles are within convenient reach of the operator when facing the area to be served by the load hook, or while facing the direction of travel of the cab. The arrangement shall allow the operator a full view of the load hook in all positions.

(c)(1)(ii)

The cab shall be located to afford a minimum of 3 inches clearance from all fixed structures within its area of possible movement.

(c)(2)

Access to crane. Access to the cab and/or bridge walkway shall be by a conveniently placed fixed ladder, stairs, or platform requiring no step over any gap exceeding 12 inches. Fixed ladders shall be in conformance with the American National Standard Safety Code for Fixed Ladders, ANSI A14.3-1956, which is incorporated by reference as specified in Sec. 1910.6.

> *While this standard (American National Standard Safety Code for Fixed Ladders, ANSI A14.3-1956) is often used as a guide by crane inspectors, it also provides some guidelines for the use of ladders.*
>
> *When climbing/descending a ladder, it is the responsibility of the user to:*
> - *Face the ladder and maintain a three-point contact at all times*
> - *Not carry tools or equipment*
> - *Never jump or slide down from a ladder or climb more than one rung/step at a time*
> - *Wear footwear with heels when climbing a ladder and avoid using flat-soled shoes*
> - *Avoid using greasy or slippery gloves and/or footwear.*
>
> *Contributing factors to falls from crane ladders include, but are not limited to:*
>
> - *Improper climbing posture creating user clumsiness on the ladder*
> - *Footwear worn by the ladder user*
> - *Users age and physical condition*
> - *Haste*
> - *Sudden movement*
> - *Lack of attention during use*
> - *Ladder condition (worn or damaged).*

(c)(3)

Fire extinguisher. Carbon tetrachloride extinguishers shall not be used.

> *Section (c)(3) does not specifically require fire extinguishers in cabs, only that carbon tetrachloride extinguishers are prohibited. However, if a fire extinguisher is provided, the employer shall ensure that operators are familiar with the operation and care of the extinguisher.*

(c)(4)

Lighting. Light in the cab shall be sufficient to enable the operator to see clearly enough to perform his work.

(d) Footwalks and Ladders

(d)(1)(i)

If sufficient headroom is available on cab-operated cranes, a footwalk shall be provided on the drive side along the entire length of the bridge of all cranes having the trolley running on the top of the girders.

(d)(1)(ii)

Where footwalks are located in no case shall less than 48 inches of headroom be provided.

> *Additional requirements for the construction of footwalks can be found in 1910.179 (d)(2)(i), (ii) and (iv).*

(d)(2)(i)

Footwalks shall be of rigid construction and designed to sustain a distributed load of at least 50 pounds per square foot.

(d)(2)(ii)

Footwalks shall have a walking surface of anti-slip type.

> *Note: Wood will meet this requirement.*

(d)(2)(iv)

The inner edge shall extend at least to the line of the outside edge of the lower cover plate or flange of the girder.

> *This standard does not establish a minimum working clearance in the direction of the access to live parts on overhead and gantry cranes. With regard to the working clearance around electrical equipment on overhead and gantry cranes, requirements are satisfied if a footwalk of at least 18 inches is provided in the direction of live parts as prescribed by Section 1910.179(d)(2)(iv), and such live parts are protected by cabinets or elevation to prevent accidental contact as prescribed by Section 110-17 of the 1971 NEC.*

(d)(3)

Toeboards and handrails for footwalks. Toeboards and handrails shall be in compliance with section 1910.23 of this part.

(d)(4)

Ladders and stairways.

(d)(4)(i)

Gantry cranes shall be provided with ladders or stairways extending from the ground to the footwalk or cab platform.

(d)(4)(ii)

Stairways shall be equipped with rigid and substantial metal handrails. Walking surfaces shall be of an anti-slip type.

(d)(4)(iii)

Ladders shall be permanently and securely fastened in place and shall be constructed in compliance with 29 CFR1910.27 (A copy of 1910.27 can be found in Appendix E).

(e) Trolley Stops

(e)(1)(i)

Stops shall be provided at the limits of travel of the trolley.

(e)(1)(ii)

Stops shall be fastened to resist forces applied when contacted.

(e)(1)(iii)

A stop engaging the tread of the wheel shall be of a height at least equal to the radius of the wheel.

(e)(2)(i)

Bridge Bumpers

A crane shall be provided with bumpers or other automatic means providing equivalent effect, unless the crane travels at a slow rate of speed and has a faster deceleration rate due to the use of sleeve bearings, or is not operated near the ends of bridge and trolley travel, or is restricted to a limited distance by the nature of the crane operation and there is no hazard of striking any object in this limited distance, or is used in similar operating conditions. The bumpers shall be capable of stopping the crane (not including the lifted load) at an average rate of deceleration not to exceed 3 ft/s/s when traveling in either direction at 20 percent of the rated load speed.

(e)(2)(i)(a)

The bumpers shall have sufficient energy absorbing capacity to stop the crane when traveling at a speed of at least 40 percent of rated load speed.

(e)(2)(i)(b)

The bumper shall be so mounted that there is no direct shear on bolts.

(e)(2)(ii)

Bumpers shall be so designed and installed as to minimize parts falling from the crane in case of breakage.

(e)(3)

Trolley bumpers

(e)(3)(i)

A trolley shall be provided with bumpers or other automatic means of equivalent effect, unless the trolley travels at a slow rate of speed, or is not operated near the ends of bridge and trolley travel, or is restricted to a limited distance of the runway and there is no hazard of striking any object in this limited distance, or is used in similar operating conditions. The bumpers shall be capable of stopping the trolley (not including the lifted load) at an average rate of deceleration not to exceed 4.7 ft/s/s when traveling in either direction at one-third of the rated load speed.

(e)(3)(ii)

When more than one trolley is operated on the same bridge, each shall be equipped with bumpers or equivalent on their adjacent ends.

(e)(3)(iii)

Bumpers or equivalent shall be designed and installed to minimize parts falling from the trolley in case of age.

(e)(4)

Rail sweeps. Bridge trucks shall be equipped with sweeps which extend below the top of the rail and project in front of the truck wheels.

(e)(5)

Guards for hoisting ropes.

(e)(5)(i)

If hoisting ropes run near enough to other parts to make fouling or chafing possible, guards shall be installed to prevent this condition.

(e)(5)(ii)

A guard shall be provided to prevent contact between bridge conductors and hoisting ropes if they could come into contact.

(e)(6)

Guards for moving parts.

(e)(6)(i)

Exposed moving parts such as gears, set screws, projecting keys, chains, chain sprockets, and reciprocating components which might constitute a hazard under normal operating conditions shall be guarded.

(e)(6)(ii)

Guards shall be securely fastened.

(e)(6)(iii)

Each guard shall be capable of supporting without permanent distortion the weight of a 200-pound person unless the guard is located where it is impossible for a person to step on it.

(f) Brakes

(f)(1)

Brakes for hoists

(f)(1)(i)

Each independent hoisting unit of a crane shall be equipped with at least one self-setting brake, hereafter referred to as a holding brake, applied directly to the motor shaft or some part of the gear train.

(f)(1)(ii)

Each independent hoisting unit of a crane, except worm-geared hoists, the angle of whose worm is such as to prevent the load from accelerating in the lowering direction shall, in addition to a holding brake, be equipped with control braking means to prevent overspeeding.

(f)(2)

Holding brakes.

(f)(2)(i)

Holding brakes for hoist motors shall have not less than the following percentage of the full load hoisting torque at the point where the brake is applied.

(f)(2)(i)(a)

125 percent when used with a control braking means other than mechanical.

(f)(2)(i)(b)

100 percent when used in conjunction with a mechanical control braking means.

(f)(2)(i)(c)

100 percent each if two holding brakes are provided.

(f)(2)(ii)

Holding brakes on hoists shall have ample thermal capacity for the frequency of operation required by the service.

(f)(2)(iii)

Holding brakes on hoists shall be applied automatically when power is removed.

> *An employer would be in violation of paragraph (f)(2)(iii) if a crane were allowed to be continued in service without adjustment or repairs when either of the two following conditions existed:*
>
> 1. *The crane is equipped with holding brakes that are not designed to be applied automatically.*

2. *The crane is equipped with holding brakes that are designed to be applied automatically when the power was removed and does not do so.*

(f)(2)(iv)

Where necessary holding brakes shall be provided with adjustment means to compensate for wear.

(f)(2)(v)

The wearing surface of all holding-brake drums or discs shall be smooth.

(f)(2)(vi)

Each independent hoisting unit of a crane handling hot metal and having power control braking means shall be equipped with at least two holding brakes.

(f)(3)

Control braking means.

(f)(5)

Application of trolley brakes.

(f)(5)(i)

On cab-operated cranes with cab on trolley, a trolley brake shall be required as specified under paragraph (f)(4) of this section.

(f)(5)(ii)

A drag brake may be applied to hold the trolley in a desired position on the bridge and to eliminate creep with the power off.

(f)(6)

Application of bridge brakes.

The requirements for stopping the motion of bridge travel are provided in this paragraph, (29 CFR 1910.179(f)(6) and are as follows:

1. *The standard requires that cab-operated cranes with cab on bridge be equipped with bridge travel brakes conforming to the requirements specified at 29 CFR 1910.179(f)(4). The standards do not require that cranes equipped with cab on bridge be equipped with an automatic bridge brake. It should be noted however, that the standard at 29 CFR 1910.179(f)(4)(viii) specifies, if holding brakes (automatic brakes) are provided they shall not prohibit the use of a drift point in the control circuit. "Drift point," is defined in Appendix A.*

2. *On cab-operated cranes with cab on trolley, a bridge brake of the holding type is required. The definition provided in Appendix A for "holding brake" clarifies this requirement. A holding brake is defined to be an automatic brake that is*

activated when the power is off. OSHA interprets this to mean that, when the controller is centered in the neutral position and the power is off to the drive motor(s), an automatic bridge brake is required on cranes with cab on trolley.

3. *On all floor, remote, and pulpit-operated crane bridge drives, a brake or non-coasting mechanical drive is required. OSHA believes that this requires the brake or braking capability to automatically become effective whenever the drive components cease to be activated. An automatic bridge braking action is required on floor, remote, and pulpit-operated crane bridge drives.*

(f)(6)(i)

On cab-operated cranes with cab on bridge, a bridge brake is required as specified under paragraph (f)(4) of this section.

(f)(6)(ii)

On cab-operated cranes with cab on trolley, a bridge brake of the holding type shall be required.

(f)(6)(iii)

On all floor, remote and pulpit-operated crane bridge drives, a brake of non-coasting mechanical drive shall be provided.

(g) Electric Equipment

(g)(1)

General.

(g)(1)(i)

Wiring and equipment shall comply with subpart S of this part.

(g)(1)(ii)

The control circuit voltage shall not exceed 600 volts for ac or dc current.

(g)(1)(iii)

The voltage at pendant push-buttons shall not exceed 150 volts for ac and 300 volts for dc

(g)(1)(iv)

Where multiple conductor cable is used with a suspended pushbutton station, the station must be supported in some satisfactory manner that will protect the electrical conductors against strain.

(g)(1)(v)

Pendant control boxes shall be constructed to prevent electrical shock and shall be clearly marked for identification of functions.

(g)(2)

Equipment.

(g)(2)(i)

Electrical equipment shall be so located or enclosed that live parts will not be exposed to accidental contact under normal operating conditions.

> *Paragraph 1910.179(g)(2)(i) provides that electrical equipment shall be so located or enclosed that live parts will not be exposed to accidental contact under normal operating conditions. This would include ensuring a cover plate was installed on any switch box in the cab of the crane because with a switch box open, any electrical wiring inside would be exposed. And, with the switch box open, the crane operator could come within inches of the energized live electrical parts when he activated the switch on the box or perhaps come within a foot or two as he got into his seat. The electrical components in the box could be energized as much as 250 volts, regardless of whether the switch in the breaker box was in the "on" or "off" position, because wires passing through some boxes provide electricity to other outlets into which an operating fan or heater can be plugged, depending on the temperature, for the cab.*
>
> *While this part of the standard addresses only "live" parts, all components in any circuit box in the cab should be considered to be energized and thereby pose a potential for the operator to receive a substantial electrical shock from any exposed wiring. While testing can establish if a circuit box is currently energized, that does not establish the fact that it could be become energized later unless it is locked and tagged out.*

The following information from NIOSH [1] stresses the importance of guarding contact conductors on the bridge of overhead cranes.

> *FACE: 93SC18*
>
> *SUBJECT: Electrician Apprentice Electrocuted After Contacting a 480-Volt Conductor in South Carolina*
>
> *CAUSE: Electrocution*
>
> *SUMMARY: On June 21, 1993, a 42-year-old male electrician apprentice was electrocuted when he inadvertently contacted an unguarded, energized 480-volt overhead-crane contact conductor.*
>
> *The victim and two co-workers had just performed electrical maintenance on the hoisting motor of an overhead crane, and the crane was moved about 200 feet to access another overhead crane. The two co-workers were on top of the bridge of the crane, while the victim operated the controls to move the crane from inside the operator's cab. After the crane was stopped alongside the second overhead crane, the victim exited the cab and climbed a metal ladder attached to the cab. The ladder led to the top of the bridge where the other workers were located. About halfway up the 7-foot metal ladder, the victim apparently partially lost his footing and his grip on the ladder. As he began to fall, he extended his left arm to catch himself and his left hand contacted one conductor of the energized, three-phase, 480-volt overhead-crane contact conductors. The electrical current passed through*

the victim's body to ground through his right hand, which was grasping the metal ladder. When the victim's hard hat fell to the ground, a worker on the ground beneath the crane looked up and saw the victim dangling from the ladder. He alerted the workers on the bridge, and they moved to help the victim. One of the workers administered cardiopulmonary resuscitation (CPR), while another ran to call for medical assistance. An ambulance arrived in about 15 minutes, and paramedics continued CPR and transported the victim to a local hospital where he was pronounced dead on arrival. NIOSH investigators concluded that to prevent similar occurrences, employers should:

- *Implement Article 610-21(a) of the National Electrical Code entitled "Locating or Guarding Contact Conductors."*

- *Identify potential hazards and appropriate safety interventions in the planning phase of maintenance projects.*

- *Routinely conduct plant surveys to identify potential and/or existing hazards and develop and implement appropriate intervention measures for these hazards.[1]*

(g)(2)(ii)

Electric equipment shall be protected from dirt, grease, oil, and moisture.

(g)(2)(iii)

Guards for live parts shall be substantial and so located that they cannot be accidently deformed so as to make contact with the live parts.

(g)(3)

Controllers.

(g)(3)(i)

Cranes not equipped with spring-return controllers or momentary contact pushbuttons shall be provided with a device which will disconnect all motors from the line on failure of power and will not permit any motor to be restarted until the controller handle is brought to the "off" position, or a reset switch or button is operated.

(g)(3)(ii)

Lever operated controllers shall be provided with a notch or latch which in the "off" position prevents the handle from being inadvertently moved to the "on" position. An "off" detent or spring return arrangement is acceptable.

(g)(3)(iii)

The controller operating handle shall be located within convenient reach of the operator.

(g)(3)(iv)

As far as practicable, the movement of each controller handle shall be in the same general directions as the resultant movements of the load.

(g)(3)(v)

The control for the bridge and trolley travel shall be so located that the operator can readily face the direction of travel.

(g)(3)(vi)

For floor-operated cranes, the controller or controllers if rope operated, shall automatically return to the "off" position when released by the operator.

(g)(3)(vii)

Pushbuttons in pendant stations shall return to the "off" position when pressure is released by the crane operator.

(g)(3)(viii)

Automatic cranes shall be so designed that all motions shall fail-safe if any malfunction of operation occurs.

(g)(3)(ix)

Remote-operated cranes shall function so that if the control signal for any crane motion becomes ineffective the crane motion shall stop.

(g)(4)

Resistors.

(g)(4)(i)

Enclosures for resistors shall have openings to provide adequate ventilation, and shall be installed to prevent the accumulation of combustible matter too near to hot parts.

(g)(4)(ii)

Resistor units shall be supported so as to be as free as possible from vibration.

(g)(4)(iii)

Provision shall be made to prevent broken parts or molten metal falling upon the operator or from the crane.

(g)(5)

Switches.

(g)(5)(i)

The power supply to the runway conductors shall be controlled by a switch or circuit breaker located on a fixed structure, accessible from the floor, and arranged to be locked in the open position.

(g)(5)(ii)

On cab-operated cranes a switch or circuit breaker of the enclosed type, with provision for locking in the open position, shall be provided in the leads from the runway conductors. A means of opening this switch or circuit breaker shall be located within easy reach of the operator.

(g)(5)(iii)

On floor-operated cranes, a switch or circuit breaker of the enclosed type, with provision for locking in the open position, shall be provided in the leads from the runway conductors. This disconnect shall be mounted on the bridge or footwalk near the runway collectors. One of the following types of floor-operated disconnects shall be provided:

(g)(5)(iii)(a)

Nonconductive rope attached to the main disconnect switch.

(g)(5)(iii)(b)

An undervoltage trip for the main circuit breaker operated by an emergency stop button in the pendant pushbutton in the pendant pushbutton station.

(g)(5)(iii)(c)

A main line contactor operated by a switch or pushbutton in the pendant pushbutton station.

(g)(5)(iv)

The hoisting motion of all electric traveling cranes shall be provided with an overtravel limit switch in the hoisting direction.

(g)(5)(v)

All cranes using a lifting magnet shall have a magnet circuit switch of the enclosed type with provision for locking in the open position. Means for discharging the inductive load of the magnet shall be provided.

(g)(6)

Runway conductors. Conductors of the open type mounted on the crane runway beams or overhead shall be so located or so guarded that persons entering or leaving the cab or crane footwalk normally could not come into contact with them.

(g)(7)

Extension lamps. If a service receptacle is provided in the cab or on the bridge of cab-operated cranes, it shall be a grounded three-prong type permanent receptacle, not exceeding 300 volts.

(h) Hoisting Equipment

(h)(1)

Sheaves.

(h)(1)(i)

Sheave grooves shall be smooth and free from surface defects which could cause rope damage.

(h)(1)(ii)

Sheaves carrying ropes which can be momentarily unloaded shall be provided with close-fitting guards or other suitable devices to guide the rope back into the groove when the load is applied again.

(h)(1)(iii)

The sheaves in the bottom block shall be equipped with close-fitting guards that will prevent ropes from becoming fouled when the block is lying on the ground with ropes loose.

(h)(1)(iv)

Pockets and flanges of sheaves used with hoist chains shall be of such dimensions that the chain does not catch or bind during operation.

(h)(1)(v)

All running sheaves shall be equipped with means for lubrication. Permanently lubricated, sealed and/or shielded bearings meet this requirement.

(h)(2)

Ropes.

(h)(2)(i)

In using hoisting ropes, the crane manufacturer's recommendation shall be followed. The rated load divided by the number of parts of rope shall not exceed 20 percent of the nominal breaking strength of the rope.

(h)(2)(ii)

Socketing shall be done in the manner specified by the manufacturer of the assembly.

(h)(2)(iii)

Rope shall be secured to the drum as follows:

(h)(2)(iii)(a)

No less than two wraps of rope shall remain on the drum when the hook is in its extreme low position.

(h)(2)(iii)(b)

Rope end shall be anchored by a clamp securely attached to the drum, or by a socket arrangement approved by the crane or rope manufacturer.

(h)(2)(iv)

Eye splices. [Reserved]

(h)(2)(v)

Rope clips attached with U-bolts shall have the U-bolts on the dead or short end of the rope. Spacing and number of all types of clips shall be in accordance with the clip manufacturer's recommendation. Clips shall be drop-forged steel in all sizes manufactured commercially. When a newly installed rope has been in operation for an hour, all nuts on the clip bolts shall be re-tightened.

(h)(2)(vi)

Swaged or compressed fittings shall be applied as recommended by the rope or crane manufacturer.

(h)(2)(vii)

Wherever exposed to temperatures, at which fiber cores would be damaged, rope having an independent wire rope or wire-strand core, or other temperature-damage resistant core shall be used.

(h)(2)(viii)

Replacement rope shall be the same size, grade, and construction as the original rope furnished by the crane manufacturer, unless otherwise recommended by a wire rope manufacturer due to actual working condition requirements.

(h)(3)

Equalizers. If a load is supported by more than one part of rope, the tension in the parts shall be equalized.

(h)(4)

Hooks. Hooks shall meet the manufacturer's recommendations and shall not be overloaded.

(i) Warning Device

Warning device. Except for floor-operated cranes a gong or other effective warning signal shall be provided for each crane equipped with a power traveling mechanism.

All overhead and gantry cranes in use shall meet the applicable requirements for design, construction, installation, testing, maintenance, inspection, and operation as prescribed in the ANSI B30.2.0-1967, Safety Code for Overhead and Gantry Cranes. Within the general scope, defined in Section I, B30.2 applies to overhead and gantry cranes, including semi-gantry, cantilever gantry, wall cranes, storage bridge cranes, and others having the same fundamental characteristics. Under the ANSI definition, it is the ability of the bridge to travel because its supporting legs are mounted on rails, not the presence of a bridge or "gantry," that makes a crane a gantry crane since it is the only characteristic mentioned. This standard applies to those cranes specifically mentioned and to other cranes having the same fundamental characteristics. These cranes are grouped because they all have trolleys and similar travel characteristics, making it clear that the emphasis of the scope of the provision is on the travel characteristics of the machine.

The following excerpts from ASME B30.2a-1991 and OSHA 1910.179 for overhead and gantry cranes indicate that warning devices SHALL be provided in the case of a remote-operated crane, and SHOULD be provided in the case of a floor-operated crane "where the ability of the operator to warn persons in the path of the load is impaired." The philosophy behind the ASME and OSHA rules appears to be that an operator who is physically constrained to be in sight of both the crane and load path needs no warning device, unless the operator's ability to warn persons in the load path is impaired. The physical constraint would be a rope, pendant, or other "means suspended from the crane." Radio transmitters do not necessarily constrain the operator to be in sight of crane or load path.

This section of 1910.179(i) requires that all cranes equipped with a power traveling mechanism, "except for floor-operated cranes," shall have "a gong or other effective warning signal." This standard specifically defines a "floor-operated crane" as one "which is pendant or nonconductive rope controlled by an operator on the floor..." This exception exists because the pendant controls, in effect, constrain the operator to remain in sight of the crane and the load path. Though an operator of a radio-controlled crane may choose to remain within sight of the crane and its load path, he is not physically compelled to do so as is the case with a pendant-operated crane. Accordingly, locations may exist within the operating area of the crane from which the crane can be operated with radio controls, but where the operator's ability to warn persons in the path of the load is impaired. For this reason, such cranes should be equipped with a gong or other warning device per this OSHA standard.

(j) Inspection

(j)(1)

Inspection classification.

(j)(1)(i)

Initial inspection. Prior to initial use all new and altered cranes shall be inspected to ensure compliance with the provisions of this section.

(j)(1)(ii)

Inspection procedure for cranes in regular service is divided into two general classifications based upon the intervals at which inspection should be performed. The intervals in turn are dependent upon the nature of the critical components of the crane and the degree of their exposure to wear, deterioration, or malfunction. The two general classifications are herein designated as "frequent" and "periodic" with respective intervals between inspections as defined below:

(j)(1)(ii)(a)

Frequent inspection - Daily to monthly intervals.

(j)(1)(ii)(b)

Periodic inspection - 1 to 12-month intervals.

(j)(2)

Frequent inspection. The following items shall be inspected for defects at intervals as defined in paragraph (j)(1)(ii) of this section or as specifically indicated, including observation during operation for any defects which might appear between regular inspections. All deficiencies such as listed shall be carefully examined and determination made as to whether they constitute a safety hazard:

(j)(2)(i)

All functional operating mechanisms for maladjustment interfering with proper operation. Daily.

(j)(2)(ii)

Deterioration or leakage in lines, tanks, valves, drain pumps, and other parts of air or hydraulic systems. Daily.

(j)(2)(iii)

Hooks with deformation or cracks. Visual inspection daily; monthly inspection with a certification record which includes the date of inspection, the signature of the person who performed the inspection and the serial number, or other identifier, of the hook inspected. For hooks with cracks or having more than 15 percent in excess of normal throat opening or more than 10^0 twist from the plane of the unbent hook refer to paragraph (l)(3)(iii)(a) of this section.

With respect to hooks, 1910.179(j)(2) means that all hooks are subject to visual inspection by the operator or other designated person(s) at daily to monthly intervals. The daily to monthly inspection intervals for crane hooks shall be determined by the frequent and periodic inspection criteria of Section 2-2.1.1 through 2-2.1.3 of ANSI/ASME B30.2-1990. Inspection of hooks found to be deformed or cracked shall comply with 1910.179(j)(2)(iii).

OSHA standard 1910.179(j)(2)(iii) and ANSI standard B30.2.0-1967, on which the OSHA standard is based, contain identical language that resulted in confusion as to the intent of the standards. Subsequent ANSI/ASME revisions to the 1967 standard (ANSI/ASME B30.2-1983/87 and 1990) resulted in further development and clarification, while the OSHA standard has not yet been modified.

Standards 1910.179(j)(2)(iii) and (iv) require "monthly inspection with a certification record which includes the date of inspection, the signature of the person who performed the inspection, and . ." With the advent of computers, some companies now use an electronic system to plan, schedule, and track equipment maintenance activities. If an employer uses an electronic system that includes all of the required information specified in the standard, it would meet the intent of the record keeping requirements. In lieu of a signature as an acknowledgment that the person delegated responsibility for the inspection, the system must identify the person who performed the inspection by certifying that it is true and complete to the best of that person's knowledge. In addition, an electronic system monthly inspection record which does not include the person's signature as required in 1910.179 would be considered a de minimis violation. De minimis violations are violations of standards which have no direct or immediate relationship to safety and health and will not be included in citations.

(j)(2)(iv)

Hoist chains, including end connections, for excessive wear, twist, distorted links interfering with proper function, or stretch beyond manufacturer's recommendations. Visual inspection daily; monthly inspection with a certification record which includes the date of inspection, the signature of the person who performed the inspection and an identifier of the chain which was inspected.

(j)(2)(vi)

All functional operating mechanisms for excessive wear of components.

(j)(2)(vii)

Rope reeving for noncompliance with manufacturer's recommendations.

(j)(3)

Periodic inspection. Complete inspections of the crane shall be performed at intervals as generally defined in paragraph (j)(1)(ii)(b) of this section, depending upon its activity, severity of service, and environment, or as specifically indicated below. These inspections shall include the requirements of paragraph (j)(2) of this section and in addition, the following items. Any deficiencies such as listed shall be carefully examined and determination made as to whether they constitute a safety hazard:

(j)(3)(i)

Deformed, cracked, or corroded members.

(j)(3)(ii)

Loose bolts or rivets.

(j)(3)(iii)

Cracked or worn sheaves and drums.

(j)(3)(iv)

Worn, cracked or distorted parts such as pins, bearings, shafts, gears, rollers, locking and clamping devices.

(j)(3)(v)

Excessive wear on brake system parts, linings, pawls, and ratchets.

(j)(3)(vi)

Load, wind, and other indicators over their full range, for any significant inaccuracies.

(j)(3)(vii)

Gasoline, diesel, electric, or other powerplants for improper performance or noncompliance with applicable safety requirements.

(j)(3)(viii)

Excessive wear of chain drive sprockets and excessive chain stretch.

(j)(3)(x)

Electrical apparatus, for signs of pitting or any deterioration of controller contactors, limit switches and pushbutton stations.

(j)(4)

Cranes not in regular use.

(j)(4)(i)

A crane which has been idle for a period of 1 month or more, but less than 6 months, shall be given an inspection conforming with requirements of paragraph (j)(2) of this section and paragraph (m)(2) of this section before placing in service.

(j)(4)(ii)

A crane which has been idle for a period of over 6 months shall be given a complete inspection conforming with requirements of paragraphs (j)(2) and (3) of this section and paragraph (m)(2) of this section before placing in service.

(j)(4)(iii)

Standby cranes shall be inspected at least semi-annually in accordance with requirements of paragraph (j)(2) of this section and paragraph (m)(2) of this section.

(k) Testing

(k)(1)(i)

Operational tests.

Prior to initial use all new and altered cranes shall be tested to insure compliance with this section including the following functions:

(k)(1)(i)(a)

Hoisting and lowering.

(k)(1)(i)(b)

Trolley travel.

(k)(1)(i)(c)

Bridge travel.

(k)(1)(i)(d)

Limit switches, locking and safety devices.

(k)(1)(ii)

The trip setting of hoist limit switches shall be determined by tests with an empty hook traveling in increasing speeds up to the maximum speed. The actuating mechanism of the limit switch shall be located so that it will trip the switch, under all conditions, in sufficient time to prevent contact of the hook or hook block with any part of the trolley.

(k)(2)

Rated load test. Test loads shall not be more than 125 percent of the rated load unless otherwise recommended by the manufacturer. The test reports shall be placed on file where readily available to appointed personnel.

It is acceptable to OSHA and within the intent of paragraph 1910.179(k)(2) that a crane may be used for rated load testing of a "below-the-hook lifting device" provided the test does not exceed 125% of the crane's rated load. Use of the crane for rated load testing of a "below the hook lifting device" may be undertaken in compliance with 1910.179(k) as specified below.

- *As often as necessary provided that the total weight on the hook does not exceed 100% of the rated load of the crane, or*

- *No more than once between each periodic crane inspection as required by1910.179(j)(1)(ii)(b) and 1910.179(j)(3) when the total weight on the hook exceeds 100% but does not exceed 125% of rated load of the crane.*

OSHA highly recommends using other acceptable methods such as higher rated cranes that will be within their rated capacity when testing below-the-hook lifting devices.

The last sentence of this paragraph reads, "the test reports shall be placed on file where readily available to appointed personnel." In order to have a record of a test for rated load and to mark the rated load certified on the crane, both would be dependent on the testing of the crane.

Also, it is noted that to be in compliance with 29 CFR 1910.179(k)(2), test loads of the hook for any given overhead crane shall not be more than 125% of the rated load of the crane unless otherwise recommended by the manufacturer.

(l) Preventive Maintenance

(l)(1)

Preventive maintenance. A preventive maintenance program based on the crane manufacturer's recommendations shall be established.

(l)(2)

Maintenance procedure.

(l)(2)(i)

Before adjustments and repairs are started on a crane the following precautions shall be taken:

(l)(2)(i)(a)

The crane to be repaired shall be run to a location where it will cause the least interference with other cranes and operations in the area.

(l)(2)(i)(b)

All controllers shall be at the off position.

(l)(2)(i)(c)

The main or emergency switch shall be open and locked in the open position.

(l)(2)(i)(d)

Warning or "out of order" signs shall be placed on the crane, also on the floor beneath or on the hook where visible from the floor.

(l)(2)(i)(e)

Where other cranes are in operation on the same runway, rail stops or other suitable means shall be provided to prevent interference with the idle crane.

(l)(2)(ii)

After adjustments and repairs have been made the crane shall not be operated until all guards have been reinstalled, safety devices reactivated and maintenance equipment removed.

(l)(3)

Adjustments and repairs.

(l)(3)(i)

Any unsafe conditions disclosed by the inspection requirements of paragraph (j) of this section shall be corrected before operation of the crane is resumed. Adjustments and repairs shall be done only by designated personnel.

(l)(3)(ii)

Adjustments shall be maintained to assure correct functioning of components. The following are examples:

(l)(3)(ii)(a)

All functional operating mechanisms.

(l)(3)(ii)(b)

Limit switches.

(l)(3)(ii)(c)

Control systems.

(l)(3)(ii)(d)

Brakes.

(l)(3)(ii)(e)

Power plants.

(l)(3)(iii)

Repairs or replacements shall be provided promptly as needed for safe operation. The following are examples:

(l)(3)(iii)(a)

Crane hooks showing defects described in paragraph (j)(2)(iii) of this section shall be discarded. Repairs by welding or reshaping are not generally recommended. If such repairs are attempted they shall only be done under competent supervision and the hook shall be tested to the load requirements of paragraph (k)(2) of this section before further use.

(l)(3)(iii)(b)

Load attachment chains and rope slings showing defects described in paragraph (j)(2)(iv) and (v) of this section respectively.

(l)(3)(iii)(c)

All critical parts which are cracked, broken, bent, or excessively worn.

(l)(3)(iii)(d)

Pendant control stations shall be kept clean and function labels kept legible.

(m) Rope Inspection

(m)(1)

Running ropes. A thorough inspection of all ropes shall be made at least once a month and a certification record which includes the date of inspection, the signature of the person who performed the inspection and an identifier for the ropes which were inspected shall be kept on file where readily available to appointed personnel. Any deterioration, resulting in appreciable loss of original strength, shall be carefully observed and determination made as to whether further use of the rope would constitute a safety hazard. Some of the conditions that could result in an appreciable loss of strength are the following:

> *Paragraph 29 CFR 179(m)(1) requires that a thorough inspection of all running ropes shall be made at least once a month. Any deterioration resulting in appreciable loss of original strength shall be carefully observed and determination made as to whether further use of the rope would constitute a safety hazard. Some of the conditions that could result in an appreciable loss of strength are the following:*
>
> - *Reduction of rope diameter below nominal diameter due to loss of core support, internal or external corrosion, or wear of outside wires*
> - *A number of broken outside wires and the degree of distribution or concentration of such broken wires*
> - *Worn outside wires*
> - *Corroded or broken wires at end connections*
> - *Corroded, cracked, bent, worn, or improperly applied end connections*
> - *Severe kinking, crushing, cutting, or unstranding.*
>
> *The standard at 29 CFR 1910.179(m)(1) also requires that a certification record be made for each wire rope inspection. The certification record must include the following:*
>
> - *The date of inspection*
> - *The signature of the person who performed the inspection*
> - *An identifier for the ropes which were inspected.*
>
> *The certification record must be kept on file where readily available to appointed personnel.*

(m)(1)(i)

Reduction of rope diameter below nominal diameter due to loss of core support, internal or external corrosion, or wear of outside wires.

(m)(1)(ii)

A number of broken outside wires and the degree of distribution or concentration of such broken wires.

(m)(1)(iii)

Worn outside wires.

(m)(1)(iv)

Corroded or broken wires at end connections.

(m)(1)(v)

Corroded, cracked, bent, worn, or improperly applied end connections.

(m)(1)(vi)

Severe kinking, crushing, cutting, or unstranding.

(m)(2)

Other ropes. All rope which has been idle for a period of a month or more due to shutdown or storage of a crane on which it is installed shall be given a thorough inspection before it is used. This inspection shall be for all types of deterioration and shall be performed by an appointed person whose approval shall be required for further use of the rope. A certification record shall be available for inspection which includes the date of inspection, the signature of the person who performed the inspection and an identifier for the rope which was inspected.

(n) Handling the Load

(n)(1)

Size of load. The crane shall not be loaded beyond its rated load except for test purposes as provided in paragraph (k) of this section.

> *If the weight of the load to be lifted is not known, either a load-indicating device must be on the crane to determine an accurate reading before the lift is attempted or an effort must be made to calculate the load as accurately as possible through engineering calculations. Load indicators have been on the market for a number of years and are quite common and accurate. They can be installed on cranes and materially reduce the risk of overlifts by reducing the problem of inaccurate or unknown weights that crane operators are often faced with.*

(n)(2)

Attaching the load.

(n)(2)(i)

The hoist chain or hoist rope shall be free from kinks or twists and shall not be wrapped around the load.

(n)(2)(ii)

The load shall be attached to the load block hook by means of slings or other approved devices.

(n)(2)(iii)

Care shall be taken to make certain that the sling clears all obstacles.

(n)(3)

Moving the load.

(n)(3)(i)

The load shall be well secured and properly balanced in the sling or lifting device before it is lifted more than a few inches.

(n)(3)(ii)

Before starting to hoist the following conditions shall be noted:

(n)(3)(ii)(a)

 Hoist rope shall not be kinked.

(n)(3)(ii)(b)

Multiple part lines shall not be twisted around each other.

(n)(3)(ii)(c)

The hook shall be brought over the load in such a manner as to prevent swinging.

(n)(3)(iii)

During hoisting care shall be taken that:

(n)(3)(iii)(a)

There is no sudden acceleration or deceleration of the moving load.

(n)(3)(iii)(b)

The load does not contact any obstructions.

(n)(3)(iv)

Cranes shall not be used for side pulls except when specifically authorized by a responsible person who has determined that the stability of the crane is not thereby endangered and that various parts of the crane will not be overstressed.

(n)(3)(v)

While any employee is on the load or hook, there shall be no hoisting, lowering, or traveling.

(n)(3)(vi)

The employer shall require that the operator avoid carrying loads over people.

(n)(3)(vii)

The operator shall test the brakes each time a load approaching the rated load is handled. The brakes shall be tested by raising the load a few inches and applying the brakes.

(n)(3)(viii)

The load shall not be lowered below the point where less than two full wraps of rope remain on the hoisting drum.

(n)(3)(ix)

When two or more cranes are used to lift a load one qualified responsible person shall be in charge of the operation. He shall analyze the operation and instruct all personnel involved in the proper positioning, rigging of the load, and the movements to be made.

(n)(3)(x)

The employer shall insure that the operator does not leave his position at the controls while the load is suspended.

(n)(3)(xi)

When starting the bridge and when the load or hook approaches near or over personnel, the warning signal shall be sounded.

(n)(4)

Hoist limit switch.

(n)(4)(i)

At the beginning of each operator's shift, the upper limit switch of each hoist shall be tried out under no load. Extreme care shall be exercised; the block shall be "inched" into the limit or run in at slow speed. If the switch does not operate properly, the appointed person shall be immediately notified.

> *Paragraph 29 CFR 1910.179(n)(4)(i) pertains to hoists that have two upper "hoist limit switches" and requires that at the beginning of each operator's shift the upper limit switch of each hoist must be tried out under no load. When the hoist is equipped with two upper hoist limit switches to prevent two-blocking, the first upper limit switch to be activated is the only upper hoist limit switch that needs to be tried at the beginning of each operator's shift. OSHA does not require a backup hoist limit switch.*

(n)(4)(ii)

The hoist limit switch which controls the upper limit of travel of the load block shall never be used as an operating control.

(o) Other Requirements, General

(o)(1)

Ladders.

(o)(1)(i)

The employer shall insure that hands are free from encumbrances while personnel are using ladders.

(o)(1)(ii)

Articles which are too large to be carried in pockets or belts shall be lifted and lowered by hand line.

(o)(2)

Cabs.

(o)(2)(i)

Necessary clothing and personal belongings shall be stored in such a manner as not to interfere with access or operation.

(o)(2)(ii)

Tools, oil cans, waste, extra fuses, and other necessary articles shall be stored in the tool box, and shall not be permitted to lie loose in or about the cab.

(o)(3)

Fire extinguishers. The employer shall insure that operators are familiar with the operation and care of fire extinguishers provided.

[39 FR 23502, June 27, 1974, as amended at 40 FR 27400, June 27, 1975; 49 FR 5322, Feb. 10, 1984; 51 FR 34560, Sept. 29, 1986; 55 FR 32015, Aug. 6, 1990; 61 FR 9227, March 7, 1996]]

1. NIOSH [1993c]: On June 21, 1993, a 42-year-old male electrician apprentice was electrocuted when he inadvertently contacted an unguarded, energized 480-volt overhead-crane contact conductor.–West Virginia. Morgantown, WV: U.S. Department of Health and Human Services, Public Health Service, Centers for Disease Control and Prevention, National Institute for Occupational Safety and Health, Fatality Assessment and Control Evaluation (FACE) Report No. 93-SC18.

> *The question is often asked whether overhead bridge cranes and gantry cranes, regulated by OSHA Standard 29 CFR 1910.179, can be used as work platforms. This would include work, such as servicing of overhead lights, that is often performed from such cranes. The following letter from OSHA says such work can be performed if certain conditions are met. An assessment checklist, "Use of an Overhead Bridge Crane as a Work Platform," located in Appendix M.5. can be used to verify that all required conditions are met.*

OSHA Standards Interpretation and Compliance Letter

04/06/1993 - Use of overhead bridge and gantry cranes as work platforms.

OSHA Standard Interpretation and Compliance Letters - Table of Contents

> *Record Type: Interpretation*
> *Standard Number: 1910.179*
> *Subject: Use of overhead bridge and gantry cranes as work platforms.*
> *Information Date:04/06/1993*

April 6, 1993

Colin J. Brigham, CIH, CSP
Loss Control Manager
EBI Companies
Four Greenwood Square
3325 Street Road
Suite 220
Bensalem, PA 19020

Dear Mr. Brigham:

This is in response to your question concerning whether overhead bridge cranes and gantry cranes, regulated by OSHA Standard 29 CFR 1910.179, can be used as work platforms. Work, such as servicing of overhead lights, would be performed from such cranes.

Work cannot be performed from a suspended load or hook but otherwise may be done from the types of cranes referenced in the preceding paragraph. However, such work may be performed only if all the following conditions are met:

1. Fall protection and standard guardrails shall be provided and used.

2. Ladders shall not be erected and/or used on cranes to gain access to areas that are not directly accessible from a crane without the use of a ladder.

3. Only "qualified persons," as that term is defined at 29 CFR 1910.399, shall be permitted to perform work on or near electrical equipment.

4. Machinery and live electrical equipment shall be guarded.

5. Work from cranes shall be performed only when a crane is stationary.

6. Crane operators shall be notified before work is performed from a crane.

7. A crane shall not be moved until all employees on the crane are in locations where they will not be exposed to injury.

8. *Crane operators shall not move a crane unless they have determined that all personnel are located in areas where they will not be exposed to injury.*

9. *Rail stops or other suitable methods shall be used to prevent a crane from being struck, whenever other cranes are in operation on the same runway.*

10. *Signs, which indicate work is being performed, shall be posted whenever work is performed from a crane. Such signs shall be visible from the floor.*

11. *Lockout/Tagout procedures shall be implemented, as appropriate, when work is being performed from a crane.*

12. *Safe egress to and from a crane shall be provided.*

If you require any additional information regarding the preceding, please contact John McFee of my staff at (215) 596-1201.

Sincerely,

LINDA R. ANKU

Regional Administrator

4.B. OSHA 1910.180 Crawler Locomotive and Truck Cranes

(a) "Definitions Applicable to This Section."

(a)(1)

A "crawler crane" consists of a rotating superstructure with power plant, operating machinery, and boom, mounted on a base, equipped with crawler treads for travel. Its function is to hoist and swing loads at various radii.

(a)(2)

A "locomotive crane" consists of a rotating superstructure with power-plant, operating machinery and boom, mounted on a base or car equipped for travel on railroad track. It may be self-propelled or propelled by an outside source. Its function is to hoist and swing loads at various radii.

(a)(3)

A "truck crane" consists of a rotating superstructure with powerplant, operating machinery and boom, mounted on an automotive truck equipped with a powerplant for travel. Its function is to hoist and swing loads at various radii.

(a)(4)

A "wheel mounted crane" (wagon crane) consists of a rotating superstructure with powerplant, operating machinery and boom, mounted on a base or platform equipped with axles and rubber-tired wheels for travel. The base is usually propelled by the engine in the superstructure, but it may be equipped with a separate engine controlled from the superstructure. Its function is to hoist and swing loads at various radii.

(a)(5)

An "accessory" is a secondary part or assembly of parts which contributes to the overall function and usefulness of a machine.

(a)(6)

"Appointed" means assigned specific responsibilities by the employer or the Employer's representative.

(a)(7)

"ANSI" means the American National Standards Institute.

(a)(8)

An "angle indicator" [boom] is an accessory which measures the angle of the boom to the horizontal.

(a)(9)

The "axis of rotation" is the vertical axis around which the crane superstructure rotates.

(a)(10)

"Axle" means the shaft or spindle with which or about which a wheel rotates. On truck- and wheel-mounted cranes it refers to an automotive type of axle assembly including housings, gearing, differential, bearings, and mounting appurtenances.

(a)(11)

"Axle" [bogie] means two or more automotive-type axles mounted in tandem in a frame so as to divide the load between the axles and permit vertical oscillation of the wheels.

(a)(12)

The "base" (mounting) is the traveling base or carrier on which the rotating superstructure is mounted such as a car, truck, crawlers, or wheel platform.

(a)(13)

The "boom" [crane] is a member hinged to the front of the rotating superstructure with the outer end supported by ropes leading to a gantry or A-frame and used for supporting the hoisting tackle.

(a)(14)

The "boom angle" is the angle between the longitudinal centerline of the boom and the horizontal. The boom longitudinal centerline is a straight line between the boom foot pin (heel pin) centerline and boom point sheave pin centerline.

(a)(15)

The "boom hoist" is a hoist drum and rope reeving system used to raise and lower the boom. The rope system may be all live reeving or a combination of live reeving And pendants.

(a)(16)

The "boom stop" is a device used to limit the angle of the boom at the highest position.

(a)(17)

A "brake" is a device used for retarding or stopping motion by friction or power means.

(a)(18)

A "cab" is a housing which covers the rotating superstructure machinery and/or operator's station. On truck-crane trucks a separate cab covers the driver's station.

(a)(19)

The "clutch" is a friction, electromagnetic, hydraulic, pneumatic, or positive mechanical device for engagement or disengagement of power.

(a)(20)

The "counterweight" is a weight used to supplement the weight of the machine in providing stability for lifting working loads.

(a)(21)

"Designated" means selected or assigned by the employer or the employer's representative as being qualified to perform specific duties.

(a)(22)

The "drum" is the cylindrical members around which ropes are wound for raising and lowering the load or boom.

(a)(23)

"Dynamic" (loading) means loads introduced into the machine or its components by forces in motion.

(a)(24)

The "gantry" (A-frame) is a structural frame, extending above the superstructure, to which the boom support ropes are reeved.

(a)(25)

A "jib" is an extension attached to the boom point to provide added boom length for lifting specified loads. The jib may be in line with the boom or offset to various angles.

(a)(26)

"Load" (working) means the external load, in pounds, applied to the crane, including the weight of load-attaching equipment such as load blocks, shackles, and slings.

(a)(27)

"Load block" [upper] means the assembly of hook or shackle, swivel, sheaves, pins, and frame suspended from the boom point.

(a)(28)

"Load block" [lower] means the assembly of hook or shackle, swivel, sheaves, pins, and frame suspended by the hoisting ropes.

(a)(29)

A "load hoist" is a hoist drum and rope reeving system used for hoisting and lowering loads.

(a)(30)

"Load ratings" are crane ratings in pounds established by the manufacturer in accordance with paragraph (c) of this section.

(a)(31)

"Outriggers" are extendable or fixed metal arms, attached to the mounting base, which rest on supports at the outer ends.

(a)(32)

"Rail clamp" means a tong-like metal device, mounted on a locomotive crane car, which can be connected to the track.

(a)(33)

"Reeving" means a rope system in which the rope travels around drums and sheaves.

(a)(34)

"Rope" refers to a wire rope unless otherwise specified.

(a)(35)

"Side loading" means a load applied at an angle to the vertical plane of the boom.

(a)(36)

A "standby crane" is a crane which is not in regular service but which is used Occasionally or intermittently as required.

(a)(37)

A "standing (guy) rope" is a supporting rope which maintains a constant distance between the points of attachment to the two components connected by the rope.

(a)(38)

"Structural competence" means the ability of the machine and its components to withstand the stresses imposed by applied loads.

(a)(39)

"Superstructure" means the rotating upper frame structure of the machine and the operating machinery mounted thereon.

(a)(40)

"Swing" means the rotation of the superstructure for movement of loads in a horizontal direction about the axis of rotation.

(a)(41)

"Swing mechanism" means the machinery involved in providing rotation of the superstructure.

(a)(42)

"Tackle" is an assembly of ropes and sheaves arranged for hoisting and pulling.

(a)(43)

"Transit" means the moving or transporting of a crane from one jobsite to another.

(a)(44)

"Travel" means the function of the machine moving from one location to another, on a jobsite.

(a)(45)

The "travel mechanism" is the machinery involved in providing travel.

(a)(46)

"Wheelbase" means the distance between centers of front and rear axles. For a multiple axle assembly the axle center for wheelbase measurement is taken as the midpoint of the assembly.

(a)(47)

The "whipline" (auxiliary hoist) is a separate hoist rope system of lighter load capacity and higher speed than provided by the main hoist.

(a)(48)

A "winch head" is a power driven spool for handling of loads by means of friction between fiber or wire rope and spool.

(b) General Requirements

(b)(1)

"Application." This section applies to crawler cranes, locomotive cranes, wheel mounted cranes of both truck and self-propelled wheel type, and any variations thereof which retain the same fundamental characteristics. This section includes only cranes of the above types, which are basically powered by internal combustion engines or electric motors and which utilize drums and ropes. Cranes designed for railway and automobile wreck clearances are excepted. The requirements of this section are applicable only to machines when used as lifting cranes.

> *Section 5-0.2.1.1 of the ANSI standard defines a crawler crane as "[a] crane consisting of a rotating superstructure with power plant, operating machinery, and boom, [and] mounted on a base.... Its function is to hoist and swing loads at various radii." If a machine has a rotating superstructure with a power plant and is mounted on a base and if it has been modified to use operating machinery that cranes are typically equipped with – brakes, load or hoist blocks, hooks, slings, shackles, etc.– (see sections 5-0.2.2.25 and 5-1.1.1.b.5 of ANSI B30.5-1968) then it clearly meets the criteria of a lifting crane.*

> *The use of a machine for lifting does not make it a "crane" if it does not have the configuration of a crane. But while the absence of a boom is dispositive in determining that a machine is not a crane, the presence of a boom does not, by itself, establish that a machine is a crane. Some machines explicitly excluded from coverage have booms, for example, draglines. Section 5.0-1 of the ANSI standard provides that the standard applies to crawler cranes "and any variations thereof which retain the same fundamental characteristics." "Retain" in this case means that a machine that starts as a crane and is modified, but keeps the basic characteristics of a crane, remains covered by the standard.*

(b)(2)

"New and existing equipment." All new crawler, locomotive, and truck cranes constructed and utilized on or after August 31, 1971, shall meet the design specifications of the American National Standard Safety Code for Crawler, Locomotive, and Truck Cranes, ANSI B30.5-1968, which is incorporated by reference as specified in Sec. 1910.6. Crawler, locomotive, and truck cranes constructed prior to August 31, 1971, should be modified to conform to those design specifications by February 15, 1972, unless it can be shown that the crane cannot feasibly or economically be altered and that the crane substantially complies with the requirements of this section.

> *Section 1926.550(b)(2) requires all crawler, locomotive, and truck cranes to comply with ANSI B30.5-1968, Safety Code for Crawler, Locomotive, and Truck Cranes. Section 1910.180 derives from this same ANSI standard. On August 2, 1988, section 1926.550 was modified by the addition of a paragraph (g), which codifies OSHA Instruction STD 1-11.2B. 53 Fed. Reg. 29116 (1988). Section 1910.180(h)(3)(v) has not been modified in this fashion. The instruction provides that employers will not be cited for violating section 1926.550(b)(2) by using cranes "to hoist and suspend employees on a work platform...when such action results in the least hazardous exposure to employees" – if a number of criteria are met. The criteria most important to this case require that a full-cycle operational test lift shall be made prior to lifting of employees. The platform shall carry twice the intended load during the test lift.*

(b)(3)

"Designated personnel." Only designated personnel shall be permitted to operate a crane covered by this section.

(c) Load Ratings

(c)(1)

"Load ratings - where stability governs lifting performance."

Load Charts

> *The rating charts, assignment of limitations, identification of attachment capacities, and hand signal illustrations are read by employer supervisory personnel, crane or derrick operators, and crews. The documentation is used to determine how to use the specific machine, how much it will be able to lift as assembled in one or a number of particular configurations, what rated capacity attachments may be used with the machine in a given configuration, and as reference for proper hand signals to be used for that particular crane or derrick. If not properly used, the machine would be subject to failures, endangering the employees in the immediate vicinity.*

> *New specifications would usually be obtained from the manufacturer and replaced by the employer. There would be some cases where the manufacturer is unavailable or cannot supply the specifications. In that situation, a qualified engineer should be hired to evaluate the equipment and develop the limitations. The employer would then need to post the new specifications.*

Table 4.A.1 Load Ratings

Type of crane mounting	Maximum load ratings (percent of tipping loads)
Locomotive, without outriggers:	
Booms 60 feet or less	85[1]
Booms over 60 feet	85[1]
Locomotive, using outriggers fully extended.	80
Crawler, without outriggers	75
Crawler, using outriggers fully extended	85
Truck and wheel mounted without outriggers or using outriggers fully extended	85

Footnote (1) Unless this results in less than 30,000 pound-feet net stabilizing moment about the rail, which shall be minimum with such booms.

(c)(1)(i)

The margin of stability for determination of load ratings, with booms of stipulated lengths at stipulated working radii for the various types of crane mountings, is established by taking a percentage of the loads which will produce a condition of tipping or balance with the boom in the least stable direction, relative to the mounting. The load ratings shall not exceed the following percentages for cranes, with the indicated types of mounting under conditions stipulated in paragraphs (c)(1)(ii) and (iii) of this section.

(c)(1)(ii)

The following stipulations shall govern the application of the values in paragraph (c)(1)(i) of this section for locomotive cranes:

(c)(1)(ii)(a)

Tipping with or without the use of outriggers occurs when half of the wheels farthest from the load leave the rail.

(c)(1)(ii)(b)

The crane shall be standing on track which is level within 1 percent grade.

(c)(1)(ii)(c)

Radius of the load is the horizontal distance from a projection of the axis of rotation to the rail support surface, before loading, to the center of vertical hoist line or tackle with load applied.

(c)(1)(ii)(d)

Tipping loads from which ratings are determined shall be applied under static conditions only (i.e., without dynamic effect of hoisting, lowering, or swinging).

(c)(1)(ii)(e)

The weight of all auxiliary handling devices such as hoist blocks, hooks, and slings shall be considered a part of the load rating.

(c)(1)(iii)

Stipulations governing the application of the values in paragraph (c)(1)(i) of this section for crawler, truck, and wheel-mounted cranes shall be in accordance with Crane Load-Stability Test Code, Society of Automotive Engineers (SAE) J765, which is incorporated by reference as specified in Sec. 1910.6.

(c)(1)(iv)

The effectiveness of these preceding stability factors will be influenced by such additional factors as freely suspended loads, track, wind, or ground conditions, condition and inflation of rubber tires, boom lengths, proper operating speeds for existing conditions, and, in general, careful and competent operation. All of these shall be taken into account by the user.

(c)(2)

"Load rating chart." A substantial and durable rating chart with clearly legible letters and figures shall be provided with each crane and securely fixed to the crane cab in a location easily visible to the operator while seated at his control station.

> *The crane industry is currently supplying information, including load charts, in a manner different than what was done and acceptable to OSHA in the past. Consequently, it is now OSHA's interpretation that the posting requirements of 29 CFR 1910.180(c)(2), "...securely fixed...;" 29 CFR 1926.550(a)(2), "...shall be conspicuously posted...;" and 29 CFR 1926.550(f)(1)(ii), "...securely fixed...;" are met when the required information is contained in a notebook securely attached to the interior of the crane cab, such as by the use of a lanyard. However, this interpretation does not change the need to have the relevant instructions, warnings, and load rating charts for a lift "visible to the operator" as required by 29 CFR 1910.180(c)(2), 29 CFR 1926.550(a)(2), and 29 CFR 1926.550(f)(1)(ii). When information is contained in book form, a way of complying with these provisions would be to provide a book holder designed to keep the book open to the relevant page(s), and located so as to be visible to the operator while at the control station.*

(d) Inspection Classifications

(d)(1)

"Initial inspection." Prior to initial use all new and altered cranes shall be inspected to insure compliance with provisions of this section.

(d)(2)

"Regular inspection." Inspection procedure for cranes in regular service is divided into two general classifications based upon the intervals at which inspection should be performed. The intervals in turn are dependent upon the nature of the critical components of the crane and the degree of their exposure to wear, deterioration, or malfunction. The two general classifications are herein designated as "frequent" and "periodic", with respective intervals between inspections as defined below:

(d)(2)(i)

Frequent inspection: Daily to monthly intervals.

(d)(2)(ii)

Periodic inspection: 1- to 12- month intervals, or as specifically recommended by the manufacturer.

(d)(3)

"Frequent inspection." Items such as the following shall be inspected for defects at intervals as defined in paragraph (d)(2)(i) of this section or as specifically indicated including observation during operation for any defects which might appear between regular inspections. Any deficiencies such as listed shall be carefully examined and determination made as to whether they constitute a safety hazard:

(d)(3)(i)

All control mechanisms for maladjustment interfering with proper operation: Daily.

(d)(3)(ii)

All control mechanisms for excessive wear of components and contamination by lubricants or other foreign matter.

(d)(3)(iii)

All safety devices for malfunction.

(d)(3)(iv)

Deterioration or leakage in air or hydraulic systems: Daily.

(d)(3)(v)

Crane hooks with deformations or cracks. For hooks with cracks or having more than 15 percent in excess of normal throat opening or more than 10 deg. twist from the plane of the unbent hook.

(d)(3)(vi)

Rope reeving for noncompliance with manufacturer's recommendations.

(d)(3)(vii)

Electrical apparatus for malfunctioning, signs of excessive deterioration, dirt, and moisture accumulation.

(d)(4)

"Periodic inspection." Complete inspections of the crane shall be performed at intervals as generally defined in paragraph (d)(2)(ii) of this section depending upon its activity, severity of service, and environment, or as specifically indicated below. These inspections shall include the requirements of paragraph (d)(3) of this section and in addition, items such as the following. Any deficiencies such as listed shall be carefully examined and determination made as to whether they constitute a safety hazard:

(d)(4)(i)

Deformed, cracked, or corroded members in the crane structure and boom.

(d)(4)(ii)

Loose bolts or rivets.

(d)(4)(iii)

Cracked or worn sheaves and drums.

(d)(4)(iv)

Worn, cracked, or distorted parts such as pins, bearings, shafts, gears, rollers and locking devices.

(d)(4)(v)

Excessive wear on brake and clutch system parts, linings, pawls, and ratchets.

(d)(4)(vi)

Load, boom angle, and other indicators over their full range, for any significant inaccuracies.

(d)(4)(vii)

Gasoline, diesel, electric, or other power plants for improper performance or noncompliance with safety requirements.

(d)(4)(viii)

Excessive wear of chain-drive sprockets and excessive chain stretch.

(d)(4)(ix)

Travel steering, braking, and locking devices, for malfunction.

(d)(4)(x)

Excessively worn or damaged tires.

(d)(5)

"Cranes not in regular use."

(d)(5)(i)

A crane which has been idle for a period of one month or more, but less than 6 months, shall be given an inspection conforming with requirements of paragraph (d)(3) of this section and paragraph (g)(2)(ii) of this section before placing in service.

(d)(5)(ii)

A crane which has been idle for a period of six months shall be given a complete inspection conforming with requirements of paragraphs (d)(3) and (4) of this section and paragraph (g)(2)(ii) of this section before placing in service.

(d)(5)(iii)

Standby cranes shall be inspected at least semiannually in accordance with requirements of paragraph(d)(3) of this section and paragraph (g)(2)(ii) of this section. Such cranes which are exposed to adverse environment should be inspected more frequently.

(d)(6)

"Inspection records." Certification records which include the date of inspection, the signature of the person who performed the inspection and the serial number, or other identifier, of the crane which was inspected shall be made monthly on critical items in use such as brakes, crane hooks, and ropes. This certification record shall be kept readily available.

> *OSHA regulations require that (cranes, derricks, or other material handling devices used solely in construction operations) be inspected during initial use and annually thereafter by a "competent person" or by a government or private agency recognized by the U.S. Department of Labor. The owner must also maintain a record of these inspections.*

(e) Testing

(e)(1)

"Operational tests."

(e)(1)(i)

In addition to prototype tests and quality-control measures, each new production crane shall be tested by the manufacturer to the extent necessary to insure compliance with the operational requirements of this paragraph including functions such as the following:

(e)(1)(i)(a)

Load hoisting and lowering mechanisms.

(e)(1)(i)(b)

Boom hoisting and lower mechanisms.

(e)(1)(i)(c)

Swinging mechanism.

(e)(1)(i)(d)

Travel mechanism.

(e)(1)(i)(e)

Safety devices.

(e)(1)(ii)

Where the complete production crane is not supplied by one manufacturer such tests shall be conducted at final assembly.

(e)(1)(iii)

Certified production-crane test results shall be made available.

(e)(2)

"Rated load test."

(e)(2)(i)

Written reports shall be available showing test procedures and confirming the adequacy of repairs or alterations.

(e)(2)(ii)

Test loads shall not exceed 110 percent of the rated load at any selected working radius.

(e)(2)(iii)

Where rerating is necessary:

(e)(2)(iii)(a)

Crawler, truck, and wheel-mounted cranes shall be tested in accordance with SAE Recommended Practice, Crane Load Stability Test Code J765 (April 1961).

(e)(2)(iii)(b)

Locomotive cranes shall be tested in accordance with paragraph (c)(1)(i) and (ii) of this section.

(e)(2)(iii)(c)

Rerating test report shall be readily available.

(e)(2)(iv)

No cranes shall be rerated in excess of the original load ratings unless such rating changes are approved by the crane manufacturer or final assembler.

(f) Maintenance Procedure

(f) "General." After adjustments and repairs have been made the crane shall not be operated until all guards have been reinstalled, safety devices reactivated, and maintenance equipment removed.

> *A publication of the American Iron and Steel Institute is entitled "Wire Rope Users Manual." On page 59 of that publication, Table 14, entitled "Diagnostic Guide to Common Wire Rope Abuses," lists nine types of abuse that can lead to broken wires: fatigue; tension; abrasion; cut, gouged or rough wire; torsion or twisting; mashing; corrosion; abrasion plus fatigue; and abrasion plus tension. For each type of abuse it lists the physical appearance of the broken wire. Any breaks caused by the rope's whipping through a crane's sheaves could be the result of one or more of the other types of abuse listed and would detract considerably from differentiating from breaks caused by an accident and breaks existing before an accident.*

(g) Rope Inspection

(g)(1)

"Running ropes." A thorough inspection of all ropes in use shall be made at least once a month and a certification record which includes the date of inspection, the signature of the person who performed the inspection and an identifier for the ropes shall be prepared and kept on file where readily available. All inspections shall be performed by an appointed or authorized person. Any deterioration, resulting in appreciable loss of original strength shall be carefully observed and determination made as to whether further use of the rope would constitute a safety hazard. Some of the conditions that could result in an appreciable loss of strength are the following:

(g)(1)(i)

Reduction of rope diameter below nominal diameter due to loss of core support, internal or external corrosion, or wear of outside wires.

(g)(1)(ii)

A number of broken outside wires and the degree of distribution of concentration of such broken wires.

(g)(1)(iii)

Worn outside wires.

(g)(1)(iv)

Corroded or broken wires at end connections.

(g)(1)(v)

Corroded, cracked, bent, worn, or improperly applied end connections.

(g)(1)(vi)

Severe kinking, crushing, cutting, or unstranding.

(g)(2)

"Other ropes."

(g)(2)(i)

Heavy wear and/or broken wires may occur in sections in contact with equalizer sheaves or other sheaves where rope travel is limited, or with saddles. Particular care shall be taken to inspect ropes at these locations.

(g)(2)(ii)

All rope which has been idle for a period of a month or more due to shutdown or storage of a crane on which it is installed shall be given a thorough inspection before it is used. This inspection shall be for all types of deterioration and shall be performed by an appointed or authorized person whose approval shall be required for further use of the rope. A certification record which includes the date of inspection, the signature of the person who performed the inspection, and an identifier for the rope which was inspected shall be prepared and kept readily available.

(g)(2)(iii)

Particular care shall be taken in the inspection of nonrotating rope.

(h) Handling the Load

(h)(1)

"Size of load."

(h)(1)(i)

No crane shall be loaded beyond the rated load, except for test purposes as provided in paragraph (e) of this section.

(h)(1)(ii)

When loads which are limited by structural competence rather than by stability are to be handled, it shall be ascertained that the weight of the load has been determined within plus or minus 10 percent before it is lifted.

(h)(2)

"Attaching the load."

(h)(2)(i)

The hoist rope shall not be wrapped around the load.

(h)(2)(ii)

The load shall be attached to the hook by means of slings or other approved devices.

(h)(3)

"Moving the load."

(h)(3)(i)

The employer shall assure that:

(h)(3)(i)(a)

The crane is level and where necessary blocked properly.

(h)(3)(i)(b)

The load is well secured and properly balanced in the sling or lifting device before it is lifted more than a few inches.

(h)(3)(ii)

Before starting to hoist, the following conditions shall be noted:

(h)(3)(ii)(a)

Hoist rope shall not be kinked.

(h)(3)(ii)(b)

Multiple part lines shall not be twisted around each other.

(h)(3)(ii)(c)

The hook shall be brought over the load in such a manner as to prevent swinging.

(h)(3)(iii)

During hoisting care shall be taken that:

(h)(3)(iii)(a)

There is no sudden acceleration or deceleration of the moving load.

(h)(3)(iii)(b)

The load does not contact any obstructions.

(h)(3)(iv)

Side loading of booms shall be limited to freely suspended loads. Cranes shall not be used for dragging loads sideways.

(h)(3)(v)

No hoisting, lowering, swinging, or traveling shall be done while anyone is on the load or hook.

(h)(3)(vi)

The operator should avoid carrying loads over people.

(h)(3)(vii)

On truck-mounted cranes, no loads shall be lifted over the front area except as approved by the crane manufacturer.

(h)(3)(viii)

The operator shall test the brakes each time a load approaching the rated load is handled by raising it a few inches and applying the brakes.

(h)(3)(ix)

Outriggers shall be used when the load to be handled at that particular radius exceeds the rated load without outriggers as given by the manufacturer for that crane. Where floats are used they shall be securely attached to the outriggers. Wood blocks used to support outriggers shall:

(h)(3)(ix)(a)

Be strong enough to prevent crushing.

(h)(3)(ix)(b)

Be free from defects.

(h)(3)(ix)(c)

Be of sufficient width and length to prevent shifting or toppling under load.

(h)(3)(x)

Neither the load nor the boom shall be lowered below the point where less than two full wraps of rope remain on their respective drums.

(h)(3)(xi)

Before lifting loads with locomotive cranes without using outriggers, means shall be applied to prevent the load from being carried by the truck springs.

> *No ANSI standard mandates that outriggers be used for every lift, only that they be used when the load to be handled at that particular radius exceeds the rated load without outriggers as given by the manufacturer for that particular crane. Thus, if the crane manufacturer's specifications are that lifts can be performed only with the outriggers fully deployed and the stabilizers extended and down, any lifts by the crane without outriggers and stabilizers extended would be prohibited.*
>
> *If a crane does not have a load chart for its operation without outriggers and stabilizers extended, and the only lift chart it has is for when the outriggers and stabilizers are extended, any use of the crane without all the outriggers extended would "exceed the rated load" for the crane.*

(h)(3)(xii)

When two or more cranes are used to lift one load, one designated person shall be responsible for the operation. He shall be required to analyze the operation and instruct all personnel involved in the proper positioning, rigging of the load, and the movements to be made.

(h)(3)(xiii)

In transit the following additional precautions shall be exercised:

(h)(3)(xiii)(a)

The boom shall be carried in line with the direction of motion.

(h)(3)(xiii)(b)

The superstructure shall be secured against rotation, except when negotiating turns when there is an operator in the cab or the boom is supported on a dolly.

(h)(3)(xiii)(c)

The empty hook shall be lashed or otherwise restrained so that it cannot swing freely.

(h)(3)(xiv)

Before traveling a crane with load, a designated person shall be responsible for determining and controlling safety. Decisions such as position of load, boom location, ground support, travel route, and speed of movement shall be in accord with his determinations.

(h)(3)(xv)

A crane with or without load shall not be traveled with the boom so high that it may bounce back over the cab.

(h)(3)(xvi)

When rotating the crane, sudden starts and stops shall be avoided. Rotational speed shall be such that the load does not swing out beyond the radii at which it can be controlled. A tag or restraint line shall be used when rotation of the load is hazardous.

(h)(3)(xvii)

When a crane is to be operated at a fixed radius, the boom-hoist pawl or other positive locking device shall be engaged.

(h)(3)(xviii)

Ropes shall not be handled on a winch head without the knowledge of the operator.

(h)(3)(xix)

While a winch head is being used, the operator shall be within convenient reach of the power unit control lever.

(h)(4)

"Holding the load."

(h)(4)(i)

The operator shall not be permitted to leave his position at the controls while the load is suspended.

(h)(4)(ii)

No person should be permitted to stand or pass under a load on the hook.

(h)(4)(iii)

If the load must remain suspended for any considerable length of time, the operator shall hold the drum from rotating in the lowering direction by activating the positive controllable means of the operator's station.

(i) Other Requirements

(i)(1) "Rail clamps."

Rail clamps shall not be used as a means of restraining tipping of a locomotive crane.

(i)(2) "Ballast or counterweight."

Cranes shall not be operated without the full amount of any ballast or counterweight in place as specified by the maker, but truck cranes that have dropped the ballast or counterweight may be operated temporarily with special care and only for light loads without full ballast or counterweight in place. The ballast or counterweight in place specified by the manufacturer shall not be exceeded.

(i)(3) "Cabs."

(i)(3)(i)

Necessary clothing and personal belongings shall be stored in such a manner as to not interfere with access or operation.

(i)(3)(ii)

Tools, oil cans, waste, extra fuses, and other necessary articles shall be stored in the tool box, and shall not be permitted to lie loose in or about the cab.

(i)(4) Refueling

(i)(4)(i)

Refueling with small portable containers shall be done with an approved safety type can equipped with an automatic closing cap and flame arrester. Refer to 1910.155(c)(3) for definition of approved.

(i)(4)(ii)

Machines shall not be refueled with the engine running.

(i)(5)

"Fire extinguishers."

(i)(5)(i)

A carbon dioxide, dry chemical, or equivalent fire extinguisher shall be kept in the cab or vicinity of the crane.

(i)(5)(ii)

Operating and maintenance personnel shall be made familiar with the use and care of the fire extinguishers provided.

> *The requirements cited in standard (i)(5)(1) is to maintain a fire extinguisher in all crane cab or operator stations only when the crane is in operation. During nonworking hours when the crane is not in operation and a fire hazard to employees does not exist, a fire extinguisher is not required. The intent of these standards is to protect employees from the usual fire hazards associated with crane operations. Employers are not required to maintain a fire extinguisher in the cab or vicinity of the crane during nonworking hours.*

(i)(6) "Swinging locomotive cranes."

A locomotive crane shall not be swung into a position where railway cars on an adjacent track might strike it, until it has been ascertained that cars are not being moved on the adjacent track and proper flag protection has been established.

(j) Operations Near Overhead Lines

(j)(1)

For operations near overhead electric lines, see 1910.333(c)(3).

> *For 1910.333(c)(3) see Appendix F. Also, Appendix G contains NIOSH Publication No. 95-108; "Preventing Electrocutions of Crane Operators and Crew Members Working Near Overhead Power Lines"*

[39 FR 23502, June 27, 1974, as amended at 49 FR 5323, Feb. 10, 1984; 51 FR 34561, Sept. 29, 1986; 53 FR 12122, Apr. 12, 1988; 55 FR 32015, Aug. 6, 1990; 61 FR 9227, March 7, 1996]

4.C. OSHA 1910.181 DERRICKS

Standard Number: 1910.181

Standard Title: Derricks.

SubPart Number: N

SubPart Title: Materials Handling and Storage

(a) Definitions Applicable to This Section

(a)(1)

A "derrick" is an apparatus consisting of a mast or equivalent member held at the head by guys or braces, with or without a boom, for use with a hoisting mechanism and operating ropes.

(a)(2)

"A-frame derrick" means a derrick in which the boom is hinged from a cross member between the bottom ends of two upright members spread apart at the lower ends and joined at the top; the boom point secured to the junction of the side members, and the side members are braced or guyed from this junction point.

(a)(3)

A "basket derrick" is a derrick without a boom, similar to a gin pole, with its base supported by ropes attached to corner posts or other parts of the structure. The base is at a lower elevation than its supports. The location of the base of a basket derrick can be changed by varying the length of the rope supports. The top of the pole is secured with multiple reeved guys to position the top of the pole to the desired location by varying the length of the upper guy lines. The load is raised and lowered by ropes through a sheave or block secured to the top of the pole.

(a)(4)

"Breast derrick" means a derrick without boom. The mast consists of two side members spread farther apart at the base than at the top and tied together at top and bottom by rigid members. The mast is prevented from tipping forward by guys

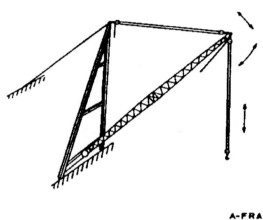

A-FRAME

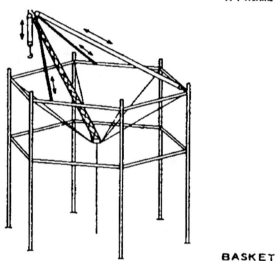

BASKET

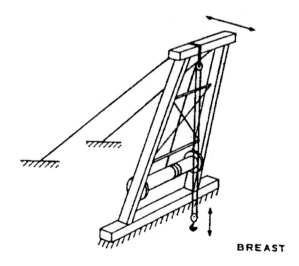

BREAST

Figure 5.A.1 Derrick Types.

connected to its top. The load is raised and lowered by ropes through a sheave or block secured to the top crosspiece.

(a)(5)

"Chicago boom derrick" means a boom which is attached to a structure, an outside upright member of the structure serving as the mast, and the boom being stepped in a fixed socket clamped to the upright. The derrick is complete with load, boom, and boom point swing line falls.

(a)(6)

A "gin pole derrick" is a derrick without a boom. Its guys are so arranged from its top as to permit leaning the mast in any direction. The load is raised and lowered by ropes reeved through sheaves or blocks at the top of the mast.

(a)(7)

"Guy derrick" means a fixed derrick consisting of a mast capable of being rotated, supported in a vertical position by guys, and a boom whose bottom end is hinged or pivoted to move in a vertical plane with a reeved rope between the head of the mast and the boom point for raising and lowering the boom, and a reeved rope from the boom point for raising and lowering the load.

(a)(8)

"Shearleg derrick" means a derrick without a boom and similar to a breast derrick. The mast, wide at the bottom and narrow at the top, is hinged at the bottom and has its top secured by a multiple reeved guy to permit handling loads at various radii by means of load tackle suspended from the mast top.

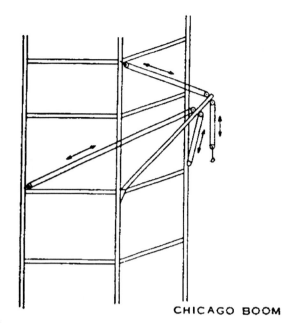

CHICAGO BOOM

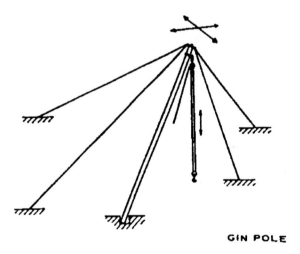

GIN POLE

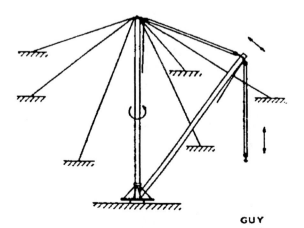

GUY

(a)(9)

A "stiffleg derrick" is a derrick similar to a guy derrick except that the mast is supported or held in place by two or more stiff members, called stifflegs, which are capable of resisting either tensile or compressive forces. Sills are generally provided to connect the lower ends of the stifflegs to the foot of the mast.

(a)(10)

"Appointed" means assigned specific responsibilities by the employer or the employer's representative.

(a)(11)

"ANSI" means the American National Standards Institute.

(a)(12)

A boom is a timber or metal section or strut, pivoted or hinged at the heel (lower end) at a location fixed in height on a frame or mast or vertical member, and with its point (upper end) supported by chains, ropes, or rods to the upper end of the frame, mast, or vertical member. A rope for raising and lowering the load is reeved through sheaves or a block at the boom point. The length of the boom shall be taken as the straight line distance between the axis of the foot pin and the axis of the boom point sheave pin, or where used, the axis of the upper load block attachment pin.

(a)(13)

"Boom harness" means the block and sheave arrangement on the boom point to which the topping lift cable is reeved for lowering and raising the boom.

(a)(14)

The "boom point" is the outward end of the top section of the boom.

(a)(15)

"Derrick bullwheel" means a horizontal ring or wheel, fastened to the foot of a derrick, for the purpose of turning the derrick by means of ropes leading from this wheel to a powered drum.

(a)(16)

"Designated" means selected or assigned by the employer or employer's representative as being qualified to perform specific duties.

(a)(17)

"Eye" means a loop formed at the end of a rope by securing the dead end to the live end at the base of the loop.

(a)(18)

A "fiddle block" is a block consisting of two sheaves in the same plane held in place by the same cheek plates.

(a)(19)

The "foot bearing" or "foot block" (sill block) is the lower support on which the mast rotates.

(a)(20)

A "gudgeon pin" is a pin connecting the mast cap to the mast allowing rotation of the mast.

(a)(21)

A "guy" is a rope used to steady or secure the mast or other member in the desired position.

(a)(22)

"Load, working" means the external load, in pounds, applied to the derrick, including the weight of load attaching equipment such as load blocks, shackles, and slings.

(a)(23)

"Load block, lower" means the assembly of sheaves, pins, and frame suspended by the hoisting rope.

(a)(24)

"Load block, upper" means the assembly of sheaves, pins, and frame suspended from the boom.

(a)(25)

"Mast" means the upright member of the derrick.

(a)(26)

"Mast cap (spider)" means the fitting at the top of the mast to which the guys are connected.

(a)(27)

"Reeving" means a rope system in which the rope travels around drums and sheaves.

(a)(28)

"Rope" refers to wire rope unless otherwise specified.

(a)(29)

"Safety Hook" means a hook with a latch to prevent slings or load from accidentally slipping off the hook.

(a)(30)

"Side loading" is a load applied at an angle to the vertical plane of the boom.

(a)(31)

The "sill" is a member connecting the foot block and stiffleg or a member connecting the lower ends of a double member mast.

(a)(32)

A "standby derrick" is a derrick not in regular service which is used occasionally or intermittently as required.

(a)(33)

"Stiffleg" means a rigid member supporting the mast at the head.

(a)(34)

"Swing" means rotation of the mast and/or boom for movements of loads in a horizontal direction about the axis of rotation.

(b) General Requirements

(b)(1)

"Application." This section applies to guy, stiffleg, basket, breast, gin pole, Chicago boom and A-frame derricks of the stationary type, capable of handling loads at variable reaches and powered by hoists through systems of rope reeving, used to perform lifting hook work, single or multiple line bucket work, grab, grapple, and magnet work. Derricks may be permanently installed for temporary use as in construction work. The requirements of this section also apply to any modification of these types which retain their fundamental features, except for floating derricks.

(b)(2)

"New and existing equipment." All new derricks constructed and installed on or after August 31, 1971, shall meet the design specifications of the American National Standard Safety Code for Derricks, ANSI B30.6-1969, which is incorporated by reference as specified in Sec. 1910.6.

(b)(3)

"Designated personnel." Only designated personnel shall be permitted to operate a derrick covered by this section.

(c) Load Ratings

(c)(1)

"Rated load marking." For permanently installed derricks with fixed lengths of boom, guy, and mast, a substantial, durable, and clearly legible rating chart shall be provided with each derrick and securely affixed where it is visible to personnel responsible for the safe operation of the equipment. The chart shall include the following data:

(c)(1)(i)

Manufacturer's approved load ratings at corresponding ranges of boom angle or operating radii.

(c)(1)(ii)

Specific lengths of components on which the load ratings are based.

(c)(1)(iii)

Required parts for hoist reeving. Size and construction of rope may be shown either on the rating chart or in the operating manual.

(c)(2)

"Nonpermanent installations." For nonpermanent installations, the manufacturer shall provide sufficient information from which capacity charts can be prepared for the particular installation. The capacity charts shall be located at the derricks or the jobsite office.

(d) Inspection

(d)(1)

"Inspection classification."

(d)(1)(i)

Prior to initial use all new and altered derricks shall be inspected to insure compliance with the provisions of this section.

(d)(1)(ii)

Inspection procedure for derricks in regular service is divided into two general classifications based upon the intervals at which inspection should be performed. The intervals in turn are dependent upon the nature of the critical components of the derrick and the degree of their exposure to wear, deterioration, or malfunction. The two general classifications are herein designated as frequent and periodic with respective intervals between inspections as defined below:

(d)(1)(ii)(a)

Frequent inspection - Daily to monthly intervals.

(d)(1)(ii)(b)

Periodic inspection 1 to 12 month intervals, or as specified by the manufacturer.

(d)(2)

"Frequent inspection." Items such as the following shall be inspected for defects at intervals as defined in paragraph (d)(1)(ii)(a) of this section or as specifically indicated, including observation during operation for any defects which might appear between regular inspections. Deficiencies shall be carefully examined for any safety hazard:

(d)(2)(i)

All control mechanisms: Inspect daily for adjustment, wear, and lubrication.

(d)(2)(ii)

All chords and lacing: Inspect daily, visually.

(d)(2)(iii)

Tension in guys: Daily.

(d)(2)(iv)

Plumb of the mast.

(d)(2)(v)

Deterioration or leakage in air or hydraulic systems: Daily.

(d)(2)(vi)

Derrick hooks for deformations or cracks; for hooks with cracks or having more than 15 percent in excess of normal throat opening or more than 10 degree twist from the plane of the unbent hook, refer to paragraph (e)(3)(iii) of this section.

(d)(2)(vii)

Rope reeving; visual inspection for noncompliance with derrick manufacturer's recommendations.

(d)(2)(viii)

Hoist brakes, clutches, and operating levers: check daily for proper functioning before beginning operations.

(d)(2)(ix)

Electrical apparatus for malfunctioning, signs of excessive deterioration, dirt, and moisture accumulation.

(d)(3)

"Periodic inspection."

(d)(3)(i)

Complete inspections of the derrick shall be performed at intervals as generally defined in paragraph (d)(1)(ii)(b) of this section depending upon its activity, severity of service, and environment, or as specifically indicated below. These inspections shall include the requirements of paragraph (d)(2) of this section and in addition, items such as the following. Deficiencies shall be carefully examined and a determination made as to whether they constitute a safety hazard:

(d)(3)(i)(a)

Structural members for deformations, cracks, and corrosion.

(d)(3)(i)(b)

Bolts or rivets for tightness.

(d)(3)(i)(c)

Parts such as pins, bearings, shafts, gears, sheaves, drums, rollers, locking and clamping devices, for wear, cracks, and distortion.

(d)(3)(i)(d)

Gudgeon pin for cracks, wear, and distortion each time the derrick is to be erected.

(d)(3)(i)(e)

Powerplants for proper performance and compliance with applicable safety requirements.

(d)(3)(i)(f)

Hooks.

(d)(3)(ii)

Foundation or supports shall be inspected for continued ability to sustain the imposed loads.

(d)(4)

"Derricks not in regular use."

(d)(4)(i)

A derrick which has been idle for a period of 1 month or more, but less than 6 months, shall be given an inspection conforming with requirements of paragraph (d)(2) of this section and paragraph (g)(3) of this section before placing in service.

(d)(4)(ii)

A derrick which has been idle for a period of over 6 months shall be given a complete inspection conforming with requirements of paragraphs (d)(2) and (3) of this section and paragraph (g)(3) of this section before placing in service.

(d)(4)(iii)

Standby derricks shall be inspected at least semiannually in accordance with requirements of paragraph (d)(2) of this section and paragraph (g)(3) of this section.

(e) Testing

(e)(1)

"Operational tests." Prior to initial use all new and altered derricks shall be tested to insure compliance with this section including the following functions:

(e)(1)(i)

Load hoisting and lowering.

(e)(1)(ii)

Boom up and down.

(e)(1)(iii)

Swing.

(e)(1)(iv)

Operation of clutches and brakes of hoist.

(e)(2)

"Anchorages." All anchorages shall be approved by the appointed person. Rock and hairpin anchorages may require special testing.

(f) Maintenance

(f)(1)

"Preventive maintenance." A preventive maintenance program based on the derrick manufacturer's recommendations shall be established.

(f)(2)

"Maintenance procedure."

(f)(2)(i)

Before adjustments and repairs are started on a derrick the following precautions shall be taken:

(f)(2)(i)(a)

The derrick to be repaired shall be arranged so it will cause the least interference with other equipment and operations in the area.

(f)(2)(i)(b)

All hoist drum dogs shall be engaged.

(f)(2)(i)(c)

The main or emergency switch shall be locked in the open position, if an electric hoist is used.

(f)(2)(i)(d)

Warning or out of order signs shall be placed on the derrick and hoist.

(f)(2)(i)(e)

The repairs of booms of derricks shall either be made when the booms are lowered and adequately supported or safely tied off.

(f)(2)(i)(f)

A good communication system shall be set up between the hoist operator and the appointed individual in charge of derrick operations before any work on the equipment is started.

(f)(2)(ii)

After adjustments and repairs have been made the derrick shall not be operated until all guards have been reinstalled, safety devices reactivated, and maintenance equipment removed.

(f)(3)

"Adjustments and repairs."

(f)(3)(i)

Any unsafe conditions disclosed by inspection shall be corrected before operation of the derrick is resumed.

(f)(3)(ii)

Adjustments shall be maintained to assure correct functioning of components.

(f)(3)(iii)

Repairs or replacements shall be provided promptly as needed for safe operation. The following are examples of conditions requiring prompt repair or replacement:

(f)(3)(iii)(a)

Hooks showing defects described in paragraph (d)(2)(vi) of this section shall be discarded.

(f)(3)(iii)(b)

All critical parts which are cracked, broken, bent, or excessively worn.

(f)(3)(iii)(c)

[Reserved]

(f)(3)(iii)(d)

All replacement and repaired parts shall have at least the original safety factor.

(g) Rope inspection

(g)(1)

"Running ropes." A thorough inspection of all ropes in use shall be made at least once a month and a certification record which includes the date of inspection, the signature of the person who performed the inspection, and an identifier for the ropes which were inspected shall be prepared and kept on file where readily available. Any deterioration, resulting in appreciable loss of original strength shall be carefully observed and determination made as to whether further use of the rope would constitute a safety hazard. Some of the conditions that could result in an appreciable loss of strength are the following:

(g)(1)(i)

Reduction of rope diameter below nominal diameter due to loss of core support, internal or external corrosion, or wear of outside wires.

(g)(1)(ii)

A number of broken outside wires and the degree of distribution or concentration of such broken wires.

(g)(1)(iii)

Worn outside wires.

(g)(1)(iv)

Corroded or broken wires at end connections.

(g)(1)(v)

Corroded, cracked, bent, worn, or improperly applied end connections.

(g)(1)(vi)

Severe kinking, crushing, cutting, or unstranding.

(g)(2)

"Limited travel ropes." Heavy wear and/or broken wires may occur in sections in contact with equalizer sheaves or other sheaves where rope travel is limited, or with saddles. Particular care shall be taken to inspect ropes at these locations.

(g)(3)

"Idle ropes." All rope which has been idle for a period of a month or more due to shutdown or storage of a derrick on which it is installed shall be given a thorough inspection before it is used. This inspection shall be for all types of deterioration. A certification record shall be prepared and kept readily available which includes the date of inspection, the signature of the person who performed the inspection, and an identifier for the ropes which were inspected.

(g)(4)

"Nonrotating ropes." Particular care shall be taken in the inspection of nonrotating rope.

(h) Operations of Derricks

Derrick operations shall be directed only by the individual specifically designated for that purpose.

(i)

"Handling the load"

(i)(1)

"Size of load."

(i)(1)(i)

No derrick shall be loaded beyond the rated load.

(i)(1)(ii)

When loads approach the maximum rating of the derrick, it shall be ascertained that the weight of the load has been determined within plus or minus 10 percent before it is lifted.

(i)(2)

"Attaching the load."

(i)(2)(i)

The hoist rope shall not be wrapped around the load.

(i)(2)(ii)

The load shall be attached to the hook by means of slings or other suitable devices.

(i)(3)

"Moving the load."

(i)(3)(i)

The load shall be well secured and properly balanced in the sling or lifting device before it is lifted more than a few inches.

(i)(3)(ii)

Before starting to hoist, the following conditions shall be noted:

(i)(3)(ii)(a)

Hoist rope shall not be kinked.

(i)(3)(ii)(b)

Multiple part lines shall not be twisted around each other.

(i)(3)(ii)(c)

The hook shall be brought over the load in such a manner as to prevent swinging.

(i)(3)(iii)

During hoisting, care shall be taken that:

(i)(3)(iii)(a)

There is no sudden acceleration or deceleration of the moving load.

(i)(3)(iii)(b)

Load does not contact any obstructions.

(i)(3)(iv)

A derrick shall not be used for side loading except when specifically authorized by a responsible person who has determined that the various structural components will not be overstressed.

(i)(3)(v)

No hoisting, lowering, or swinging shall be done while anyone is on the load or hook.

(i)(3)(vi)

The operator should avoid carrying loads over people.

(i)(3)(vii)

The operator shall test the brakes each time a load approaching the rated load is handled by raising it a few inches and applying the brakes.

(i)(3)(viii)

Neither the load nor boom shall be lowered below the point where less than two full wraps of rope remain on their respective drums.

(i)(3)(ix)

When rotating a derrick, sudden starts and stops shall be avoided. Rotational speed shall be such that the load does not swing out beyond the radius at which it can be controlled.

(i)(3)(x)

Boom and hoisting rope systems shall not be twisted.

(i)(4)

"Holding the load."

(i)(4)(i)

The operator shall not be allowed to leave his position at the controls while the load is suspended.

(i)(4)(ii)

People should not be permitted to stand or pass under a load on the hook.

(i)(4)(iii)

If the load must remain suspended for any considerable length of time, a dog, or pawl and ratchet, or other equivalent means, rather than the brake alone, shall be used to hold the load.

(i)(5)

"Use of winch heads."

(i)(5)(i)

Ropes shall not be handled on a winch head without the knowledge of the operator.

(i)(5)(ii)

While a winch head is being used, the operator shall be within convenient reach of the power unit control lever.

(i)(6)

"Securing boom." Dogs, pawls, or other positive holding mechanism on the hoist shall be engaged. When not in use, the derrick boom shall:

(i)(6)(i)

Be laid down;

(i)(6)(ii)

Be secured to a stationary member, as nearly under the head as possible, by attachment of a sling to the load block; or

(i)(6)(iii)

Be hoisted to a vertical position and secured to the mast.

(j) Other Requirements

(j)(1)

"Guards."

(j)(1)(i)

Exposed moving parts, such as gears, ropes, setscrews, projecting keys, chains, chain sprockets, and reciprocating components, which constitute a hazard under normal operating conditions shall be guarded.

(j)(1)(ii)

Guards shall be securely fastened.

(j)(1)(iii)

Each guard shall be capable of supporting without permanent distortion, the weight of a 200-pound person unless the guard is located where it is impossible for a person to step on it.

(j)(2)

"Hooks."

(j)(2)(i)

Hooks shall meet the manufacturer's recommendations and shall not be overloaded.

(j)(2)(ii)

Safety latch type hooks shall be used wherever possible.

(j)(3)

"Fire extinguishers."

(j)(3)(i)

A carbon dioxide, dry chemical, or equivalent fire extinguisher shall be kept in the immediate vicinity of the derrick.

(j)(3)(ii)

Operating and maintenance personnel shall be familiar with the use and care of the fire extinguishers provided.

(j)(4)

"Refueling."

(j)(4)(i)

Refueling with portable containers shall be done with approved safety type containers equipped with automatic closing cap and flame arrester. Refer to 1910.155(c)(3) for definition of approved.

> *1910.155(c)(3)*
>
> *"Approved" means acceptable to the Assistant Secretary under the following criteria:*
>
> *(c)(3)(i)*
>
> *If it is accepted, or certified, or listed, or labeled or otherwise determined to be safe by a nationally recognized testing laboratory; or*
>
> *(c)(3)(ii)*
>
> *With respect to an installation or equipment of a kind which no nationally recognized testing laboratory accepts, certifies, lists, labels, or determines to be safe, if it is inspected or tested by another Federal agency and found in compliance with the provisions of the applicable National Fire Protection Association Fire Code; or*
>
> *(c)(3)(iii)*
>
> *With respect to custom-made equipment or related installations which are designed, fabricated for, and intended for use by its manufacturer on the basis of test data which the employer keeps and makes available for inspection to the Assistant Secretary.*
>
> *(c)(3)(iv)*
>
> *For the purposes of paragraph (c)(3) of this section:*
>
> *(c)(3)(iv)(A)*
>
> *Equipment is listed if it is of a kind mentioned in a list which is published by a nationally recognized testing laboratory which makes periodic inspections of the production of such equipment and which states that such equipment meets nationally recognized standards or has been tested and found safe for use in a specified manner;*

(c)(3)(iv)(B)

Equipment is labeled if there is attached to it a label, symbol, or other identifying mark of a nationally recognized testing laboratory which makes periodic inspections of the production of such equipment, and whose labeling indicates compliance with nationally recognized standards or tests to determine safe use in a specified manner;

(c)(3)(iv)(C)

Equipment is accepted if it has been inspected and found by a nationally recognized testing laboratory to conform to specified plans or to procedures of applicable codes; and

(c)(3)(iv)(D)

Equipment is certified if it has been tested and found by a nationally recognized testing laboratory to meet nationally recognized standards or to be safe for use in a specified manner or is of a kind whose production is periodically inspected by a nationally recognized testing laboratory, and if it bears a label, tag, or other record of certification.

(j)(4)(ii)

Machines shall not be refueled with the engine running.

(j)(5)

"Operations near overhead lines." For operations near overhead electric lines, see 1910.333(c)(3). Appendix F)

(j)(6)

"Cab or operating enclosure."

(j)(6)(i)

Necessary clothing and personal belongings shall be stored in such a manner as to not interfere with access or operation.

(j)(6)(ii)

Tools, oilcans, waste, extra fuses, and other necessary articles shall be stored in the toolbox, and shall not be permitted to lie loose in or about the cab or operating enclosure.

[37 FR 22102, Oct. 18, 1972, as amended at 38 FR 14373, June 1, 1973; 43 FR 49750, Oct. 24, 1978; 49 FR 5323, Feb. 10, 1984; 51 FR 34561, Sept. 29, 1986; 53 FR 12122, Apr. 12, 1988; 55 FR 32015, Aug. 6, 1990; 61 FR 9227, March 7, 1996]

4.D. 1926.550 - Cranes and Derricks

Standard Number: 1926.550

Standard Title: Cranes and derricks.

Subpart Number: N

Subpart Title: Cranes, Derricks, Hoists, Elevators, and Conveyors

Background. The Final Rule, 29 CFR 1926.550, Crane and Derrick Suspended Personnel Platforms, was published in the Federal Register, Vol. 53, No. 148, on August 2, 1988.

(a) General Requirements

This standard (1926.550) sets safety requirements for the operation, maintenance, and inspection of cranes and derricks used in construction, with special requirements in subsections (b) through (f) for crawler, locomotive, and truck cranes; hammerhead tower cranes, overhead, and gantry cranes; derricks; and floating cranes and derricks.

This standard also is intended to ensure that the employer has sufficient information to institute an adequate maintenance program. OSHA does not require a routine test that exceeds the rated capacity of the crane, and overload tests are only required of new cranes and after the crane has been repaired or re-rated. Therefore, all other overload tests must be performed according to the manufacturer's specifications and limitations as required by 1926.550(a)(1).

It should be noted that OSHA construction regulations provide minimum safety and health standards for employees working in the construction industry. However, more stringent requirements, such as an ANSI standard for crane inspections at nuclear power plants, may be required by virtue of the contractual relationship for construction at the facility. In any event, employers are not precluded from obtaining the highest rating from employees to perform crane inspections.

(a)(1)

The employer shall comply with the manufacturer's specifications and limitations applicable to the operation of any and all cranes and derricks. Where manufacturer's specifications are not available, the limitations assigned to the equipment shall be based on the determinations of a qualified engineer competent in this field and such determinations will be appropriately documented and recorded. Attachments used with cranes shall not exceed the capacity, rating, or scope recommended by the manufacturer.

With regard to whether repairs can be made to a crane without the approval of the crane manufacturer, there are two provisions in 29 CFR 1926.550 that require crane users to follow procedures or gain approval of the crane manufacturer. Neither specifically addresses repairs made to the crane. Paragraph 1926.550(a)(1) requires that the employer comply with the manufacturer's specifications and limitations applicable to the operation of any and all cranes and derricks. OSHA has interpreted the language "specifications and limitations applicable to the operation" to apply to the way a crane is used and maintained. This section would not apply to a limitation on where repairs could be made unless the repair affected the safe operation of the crane.

(a)(2)

Rated load capacities, and recommended operating speeds, special hazard warnings, or instruction, shall be conspicuously posted on all equipment. Instructions or warnings shall be visible to the operator while he is at his control station.

> *There have been significant numbers of fatalities and injuries due to improper crane use. The use of appropriately documented capacity charts, machine configuration limitations, and hand signal illustrations are required to ensure the employee a safe place to work during operations involving the use of a crane or derrick. All factory-manufactured cranes have the required documentation permanently attached or included with the equipment when delivered. The manufacturer's specifications are delivered with the equipment and the load charts are permanently attached to the equipment. Without the required documentation to guide the user, cranes and derricks would be misused because information necessary for safe use would be unavailable. This documentation is used by employers, engineers, supervisors and equipment operators to determine the machine's capacity to lift and place loads and as a reference for proper hand signals to be used.*

> *These provisions require that where manufacturer's specifications are not available, the limitations assigned to a crane or derrick be based on the determinations of a qualified engineer and that load charts be posted on all equipment. The capacities of cranes and derricks are limited by the structural capacity of the components and the way those components are assembled, moved, supported, and controlled. All factory-manufactured equipment has the required documentation attached or included with the equipment as delivered. If for some reason the manufacturer's specifications are unavailable (if they have been lost or damaged), the employer must obtain a replacement set of specifications from the manufacturer. In the event the crane manufacturer is not available or cannot provide the information, the equipment must be evaluated and tested by a qualified engineer to obtain the information needed.*

(a)(4)

Hand signals to crane and derrick operators shall be those prescribed by the applicable ANSI standard for the type of crane in use. An illustration of the signals shall be posted at the job site.

> *Section 1926.550(a)(4) requires that hand signal illustrations be posted at the job site. Hand signals used during the operation of the crane or derrick are important to the safe operation of the equipment. Improper use of hand signals can lead to unsafe operation of the crane or derrick. Posting of hand signal illustrations is needed for reference at the job site.*

> *It is usual and customary for the crane or derrick cab to have a copy of hand signal illustrations attached to the equipment either inside the cab or outside on the equipment itself in the form of a decal. The decals should be inspected and replaced if necessary during the required annual crane inspection.*

(a)(5)

The employer shall designate a competent person who shall inspect all machinery and equipment prior to each use, and during use, to make sure it is in safe operating condition. Any deficiencies shall be repaired, or defective parts replaced, before continued use.

> *Section 1926.550(a)(5) requires that a competent person, designated by the employer, shall inspect all machinery for safety "prior to each use, and during use," and to make sure any deficiencies discovered during an inspection are corrected before the machine is operated again. This includes replacing any defective parts found during the inspection. An example of a violation of this standard would be the operation of a crane with one of its two boom stops broken. The absence of a boom stop is a safety defect within the meaning of this standard since with only one boom stop in place the boom can twist as it is raised, and if it is raised too far while twisting it can fall over.*

(a)(6)

A thorough, annual inspection of the hoisting machinery shall be made by a competent person, or by a government or private agency recognized by the U.S. Department of Labor. The employer shall maintain a record of the dates and results of inspections for each hoisting machine and piece of equipment.

> *Both sections 1926.550(a)(5 and (a)(6) require a thorough, annual inspection of the hoisting machinery used in construction to be made by a competent person, or by a government or private agency recognized by the U.S. Department of Labor. In addition, the employer is required to maintain a record of the dates and results of inspection for each hoisting machine and piece of equipment. The OSHA construction standards do not require employees performing crane inspections to have a Level II rating, which is a term used in an ANSI standard not referenced in the OSHA standards. This subsection, however, allows the employer to use a government or private agency recognized by the U.S. Department of Labor to perform these annual inspections. OSHA maintains that, in order to be a competent person for purposes of these standards, one must at least have had specific training in and be knowledgeable about hoisting machinery and equipment operations, manufacturer(s) specifications and recommendations, and other pertinent provisions of Subpart N-Cranes, Derricks, Hoists, Elevators, and Conveyors in 29 CFR Part 1926.*

> *For example, a competent person performing hoisting machinery and equipment inspections would be capable of identifying existing and predictable hazards such as wear, damage, corrosion, system leaks, contamination, dust accumulations, etc., and have the authorization to take prompt corrective measures. Such mea-*

sures may be repairs, part replacements, adjustments, systems disassembly, cleaning, or other activities necessary to eliminate the hazard potential surrounding the use of hoisting machinery and equipment. OSHA's position is that a person who does not have a thorough knowledge of the requirements, regulations and standards governing his/her direct duties cannot be considered as being a "competent person." This position has been upheld in a number of contested cases.

While not defined in 1926.550, 1926.32 defines many of the terms used in the construction safety and health regulations including many used in the sections pertaining to crane safety. A competent person is defined as "one who is capable of identifying existing and predictable hazards in the surroundings or working conditions which are unsanitary, hazardous, or dangerous to employees, and who has authorization to take prompt corrective measures to eliminate them". Subsection (i) in 1926.32 defines a "designated person" as meaning an "authorized person" as defined in subsection (d). Subsection (d) defines an "authorized person" as one who the employer has approved or assigned to perform a specific type of duty or duties at specific locations at the job site. If employees meet the qualifications for a "competent person" as defined in 1926.32, they can perform the construction inspections required in 1926.550.

(a)(7)

Wire rope shall be taken out of service when any of the following conditions exist:

(a)(7)(i)

In running ropes, six randomly distributed broken wires in one lay or three broken wires in one strand in one lay;

(a)(7)(ii)

Wear of one-third the original diameter of outside individual wires. Kinking, crushing, bird caging, or any other damage resulting in distortion of the rope structure;

(a)(7)(iii)

Evidence of any heat damage from any cause;

(a)(7)(iv)

Reductions from nominal diameter of more than one-sixty-fourth inch for diameters up to and including five-sixteenths inch, one-thirty-second inch for diameters three-eighths inch to and including one-half inch, three-sixty-fourths inch for diameters nine-sixteenths inch to and including three-fourths inch, one-sixteenth inch for diameters seven-eighths inch to 1 1/8 inches inclusive, three-thirty-seconds inch for diameters 1 1/4 to 1 1/2 inches inclusive;

(a)(7)(v)

In standing ropes, more than two broken wires in one lay in sections beyond end connections or more than one broken wire at an end connection.

(a)(7)(vi)

Wire rope safety factors shall be in accordance with American National Standards Institute B 30.5-1968 or SAE J959-1966.

(a)(8)

Belts, gears, shafts, pulleys, sprockets, spindles, drums, fly wheels, chains, or other reciprocating, rotating, or other moving parts or equipment shall be guarded if such parts are exposed to contact by employees, or otherwise create a hazard. Guarding shall meet the requirements of the American National Standards Institute B15.1-1958 Rev., Safety Code for Mechanical Power Transmission Apparatus.

(a)(9)

Accessible areas within the swing radius of the rear of the rotating superstructure of the crane, either permanently or temporarily mounted, shall be barricaded in such a manner as to prevent an employee from being struck or crushed by the crane.

> *Accessible areas within the swing radius of the rear of the rotating superstructure of the crane are meant to be those areas within the 360 degrees of travel. A barricade is required in accessible areas when the swing radius of the rear of the rotating structure of the crane, either permanently or temporarily mounted, may cause an employee to be crushed between other objects and the rear of the crane of any portion of the crane carriage regardless of whether it be front, rear, or side. Historically, most fatalities have been caused by employees getting between the rotating superstructure and the turntable or other portions of the crane itself.*

(a)(10)

All exhaust pipes shall be guarded or insulated in areas where contact by employees is possible in the performance of normal duties.

(a)(11)

Whenever internal combustion engine powered equipment exhausts in enclosed spaces, tests shall be made and recorded to see that employees are not exposed to unsafe concentrations of toxic gases or oxygen deficient atmospheres.

(a)(12)

All windows in cabs shall be of safety glass, or equivalent, that introduces no visible distortion that will interfere with the safe operation of the machine.

(a)(13)(i)

Where necessary for rigging or service requirements, a ladder, or steps, shall be provided to give access to a cab roof.

(a)(13)(ii)

Guardrails, handholds, and steps shall be provided on cranes for easy access to the car and cab, conforming to American National Standards Institute B30.5.

(a)(13)(iii)

Platforms and walkways shall have anti-skid surfaces.

(a)(14)

Fuel tank filler pipe shall be located in such a position, or protected in such manner, as to not allow spill or overflow to run onto the engine, exhaust, or electrical equipment of any machine being fueled.

(a)(14)(i)

An accessible fire extinguisher of 5BC rating, or higher, shall be available at all operator stations or cabs of equipment.

(a)(14)(ii)

All fuels shall be transported, stored, and handled to meet the rules of Subpart F of this part. When fuel is transported by vehicles on public highways, Department of Transportation rules contained in 49 CFR Parts 177 and 393 concerning such vehicular transportation are considered applicable.

(a)(15)

Except where electrical distribution and transmission lines have been de-energized and visibly grounded at point of work or where insulating barriers, not a part of or an attachment to the equipment or machinery, have been erected to prevent physical contact with the lines, equipment or machines shall be operated proximate to power lines only in accordance with the following:

> *OSHA considers the use of insulated gloves to be an acceptable substitute for the measures specified in (a)(15), i.e., maintaining the specified clearance, de-energizing the line, or erecting a physical barrier. OSHA would use the general duty clause of the OSH Act, 29 U.S.C.654(a)(1) or the general personal protection equipment standard at 29 CFR 1910.132(a) [Since June 30, 1993, codified within Part 1926 at 1926.95(a)] to require protective equipment.*

(a)(15)(i)

For lines rated 50 kV. or below, minimum clearance between the lines and any part of the crane or load shall be 10 feet;

> *The following guidance is provided when using insulating barriers to prevent physical contact of equipment or machinery with electric distribution and transmission lines when the equipment or machinery is operating within 10 feet of the power lines. The minimum clearance of 10 feet as used in (a)(15)(i) applies in any direc-*

tion from any applicable line. This excludes extending the minimum clearance due to the height of the lines or the height of any part of the crane or load.

Rubber insulating equipment meeting the requirements of 29 CFR 1910.137 is normally intended as protection from "brushing" type contact for employees working on the lines. Although 1910.137 is not applicable to construction work, it may be used as a compliance guide for barriers required under the exception to 1926.550(a)(15), under certain conditions. If hard direct contact with the line is not likely, rubber insulating equipment can provide protection from brush contact with the power line. However, if direct impact with the lines is reasonably likely or expected, this equipment will not provide the necessary protection. In such cases, other types of barriers would be required, such as those listed in the National Safety Council Data Sheet No. 1-743New90 and the types of plastic guard equipment covered in ASTM F968, Specification for Electrically Insulating Plastic Guard Equipment for Protection of Workers. Although guards of a type consisting of ABS plastic, 1/8-inch thick, (approximate puncture strength 50,000 volts) are often successfully used on 15KV and 34.5KV systems, none are totally impact proof to the extent that strong direct blows would leave the air gap integrity unchanged or not cause sliding or other adverse movement along the line.

While the "goal post" type of guarding approach to overhead line safety probably provides the most durable means of withstanding barrier impact, it should be remembered that no practical barrier can absolutely prevent contact of a crane (or similar material handling device), simply because the capabilities of such heavy operating devices normally overwhelm any obstruction that may be installed.

(a)(15)(ii)

For lines rated over 50 kV., minimum clearance between the lines and any part of the crane or load shall be 10 feet plus 0.4 inch for each 1 kV. over 50 kV., or twice the length of the line insulator, but never less than 10 feet;

(a)(15)(iii)

In transit with no load and boom lowered, the equipment clearance shall be a minimum of 4 feet for voltages less than 50 kV., and 10 feet for voltages over 50 kV., up to and including 345 kV., and 16 feet for voltages up to and including 750 kV.

(a)(15)(iv)

A person shall be designated to observe clearance of the equipment and give timely warning for all operations where it is difficult for the operator to maintain the desired clearance by visual means;

Regulation (a)(15)(iv) calls for the use of a spotter whenever it is difficult to obtain clearance by visual means and clearly says that a spotter "shall be designated," indicating that affirmative action must be taken by the employer. ("Designate" requires specific and positive action by the employer to inform an employee of the existence and nature of his duties).

With regard to the person designated as an observer as required by this regulation, the observer must be (1) specifically assigned by the employer, (2) the designated observer can be assigned other duties provided the other duties do not interfere with the observer's ability to give timely warning for all operations of the crane and (3) the observer must be positioned so as to be able to visually monitor the clearance between the equipment and the power lines. Although the observer's position is site dependant, it would not normally be acceptable for an observer to be positioned under the crane because the observer would have a similar visual perspective of the equipment as the crane operator.

(a)(15)(v)

Cage-type boom guards, insulating links, or proximity warning devices may be used on cranes, but the use of such devices shall not alter the requirements of any other regulation of this part even if such device is required by law or regulation;

(a)(15)(vi)

Any overhead wire shall be considered to be an energized line unless and until the person owning such line or the electrical utility authorities indicate that it is not an energized line and it has been visibly grounded;

The employer, not the crane oiler or operator, is required under section (a)(15)(vi) to make an independent inquiry to determine if overhead power lines are energized. With few exceptions, OSHA holds the employer responsible for complying with safety and health regulations by assuring that a safe distance between the crane equipment and overhead power lines is maintained or that the power lines are de-energized.

(a)(15)(vii)

Prior to work near transmitter towers where an electrical charge can be induced in the equipment or materials being handled, the transmitter shall be de-energized or tests shall be made to determine if electrical charge is induced on the crane. The following precautions shall be taken when necessary to dissipate induced voltages:

(a)(15)(vii)(a)

The equipment shall be provided with an electrical ground directly to the upper rotating structure supporting the boom; and

(a)(15)(vii)(b)

Ground jumper cables shall be attached to materials being handled by boom equipment when electrical charge is induced while working near energized transmitters. Crews shall be provided with nonconductive poles having large alligator clips or other similar protection to attach the ground cable to the load.

(a)(15)(vii)(c)

Combustible and flammable materials shall be removed from the immediate area prior to operations.

(a)(16)

No modifications or additions which affect the capacity or safe operation of the equipment shall be made by the employer without the manufacturer's written approval. If such modifications or changes are made, the capacity, operation, and maintenance instruction plates, tags, or decals, shall be changed accordingly. In no case shall the original safety factor of the equipment be reduced.

Paragraph 1926.550(a)(16) addresses the requirement for manufacturers' approval before any modifications or additions can be made which would affect the safe operation of the crane. The section goes on to address what must be done when the capacity is changed. The intent of this paragraph is to ensure that any modification or addition that would change the capacity of the crane be reviewed by the manufacturer to make sure the change does not adversely affect other crane functions or components and cause an unsafe condition. By their very nature major repairs to crane components create the potential for adversely affecting capacity or safe operation, and for that reason such repairs should be reviewed and approved by the manufacturer to make sure that the crane's capacity and safe operation are not affected. Alternatively, this could be done by seeking the certification of a registered professional engineer that a repair has restored the crane component to its original configuration and strength so that the capacity and operation of the crane is unaffected by the repair. A crane user who has had major repairs carried out but has not taken appropriate steps to ensure that capacity or safe operation has not been adversely affected could be cited by OSHA. The OSHA penalties that could come from a citation issued for violations of these provisions is like all citations issued by OSHA; penalties are assessed based on the circumstances (including violation history of employer, seriousness of hazard, number of workers exposed, and other related factors) of each case.

In regard to what replacement parts can be used on a crane without the approval of the crane manufacturer, an employer can use replacement components without the approval of the crane manufacturer if the new components are equivalent in strength and durability and do not affect the rated capacity or the safe operations of the crane. If a modification or a replacement part (i.e., boom, shaft, gear, etc.) will affect the rated capacity or safe operation, then the modification or part can only be made with the written approval of the crane manufacturer and all affected instruction plates, load charts, and other related printed information must be revised accordingly. It is important to note that placing a crane in operation without revising the load chart to reflect such a modification or part change would constitute a violation of paragraph 1926.550(a)(16). Moreover, when the manufacturer has specified that unapproved replacement parts are not to be used, OSHA may presume an adverse effect on safety if this limitation is not followed. See 1926.550(a)(1).

As mentioned above, this provision also requires that when a crane or derrick has been modified, it must be evaluated and tested by a qualified engineer to determine the capacity, operation, and maintenance instructions. Occasionally, because of job environment limitations or job-specific needs, an employer will assemble components of several manufactured mobile crane machines. Rating charts and limitation guideline documents for unitized machines sold to employers will not provide the information required for these field-modified or field-manufactured machines. The reason is that diverse components that have been substituted for unitized machine components (such as carriers, outriggers, or swing tables) will have differing strengths according to the dimensional variants of the respective components. When these components are changed, the appropriate rating for the newly matched assembly must be ascertained to ensure safe operation of the machine. For cranes or derricks that are modified in accordance with paragraph (a)(16), posting of the load charts and specifications would be required in all instances.

(a)(17)

The employer shall comply with Power Crane and Shovel Association Mobile Hydraulic Crane Standard No. 2.

It is intended that 29 CFR 1926.550 cover all truck cranes [see 1926.550(b)(2)]. However, the reference to the Power Crane and Shovel Association Mobile Hydraulic Crane Standard No. 2. of the above-mentioned reference refers the user back to ANSI B30.5 for additional safety practices and to that degree both standards are applicable to hydraulic truck cranes.

(a)(18)

Sideboom cranes mounted on wheel or crawler tractors shall meet the requirements of SAE J743a-1964.

(a)(19)

All employees shall be kept clear of loads about to be lifted and of suspended loads.

The June 30, 1993 Federal Register notice (58FR35076) incorporated certain specified general industry-standards into the construction standards. The preamble to the subject standard clearly states that "the Agency is limiting this rulemaking to the prevention of accidents in general industry and maritime ... because to include other industrial sectors (such as construction) would seriously impede the rulemaking process" (55FR31986). Further, the preamble states "the standard will apply to every major standard industrial code (SIC) economic division with the exception of ...construction...." (55FR32011).

In addition, general industry provision 29 CFR 1910.184(c)(9), which was codified as 29 CFR 1926.550(a)(19) and which reads "[all employees shall be kept clear of loads about to be lifted and of suspended loads," is not applicable to crawler, truck, or locomotive cranes used for construction activities. This is because the general industry provision is preempted from being applied to such cranes by the more specific construction rule [1926.550(b)(2)] for such cranes that references the ANSI B30.5-1968 standard. In the ANSI document, paragraph 5-3.2.3f states "the operator should avoid carrying loads over people." Because the ANSI rule is advisory it is not enforceable. Consequently, except for 1926.701(e)(2), which addresses concrete buckets, and 1926.651(e) which addresses loads handled by lifting or digging equipment, there is no OSHA provision that prohibits using crawler, truck, or locomotive cranes to suspend loads over employees.

(b) Crawler, Locomotive, and Truck Cranes.

(b)(1)

All jibs shall have positive stops to prevent their movement of more than 5 degrees above the straight line of the jib and boom on conventional type crane booms. The use of cable type belly slings does not constitute compliance with this rule.

(b)(2)

All crawler, truck, or locomotive cranes in use shall meet the applicable requirements for design, inspection, construction, testing, maintenance and operation as prescribed in the ANSI B30.5-1968, Safety Code for Crawler, Locomotive and Truck Cranes. However, the written, dated, and signed inspection reports and records of the monthly inspection of critical items prescribed in section 5-2.1.5 of the ANSI B30.5-1968 standard are not required. Instead, the employer shall prepare a certification record which includes the date the crane items were inspected; the signature of the person who inspected the crane items; and a serial number, or other identifier, for the crane inspected. The most recent certification record shall be maintained on file until a new one is prepared.

This section of the standard [29 CFR 1926.550(b)(2)] requires that the design, inspection, construction, testing, maintenance, and operation of crawler cranes meet the American National Standards Institute (ANSI) standard B30.5-1968, Safety Code for Crawler, Locomotive and Truck Cranes. Section 5-2.1.3 of the ANSI standard requires that brakes and clutch system parts, linings, pawls, and ratchets be periodically inspected for excessive wear. Section 5-2.3.3 requires that any unsafe conditions disclosed by the frequent and periodic inspection requirements be corrected before operation of the crane is resumed. One reason for this requirement is that the lack of inspection and maintenance of a crane can lead to the simultaneous failure of both the primary and secondary control systems on the crane's cable drum.

Section 1926.550(b)(2) requires all crawler, locomotive, and truck cranes to comply with ANSI B30.5-1968, Safety Code for Crawler, Locomotive and Truck Cranes. Section 1910.180 derives from this same ANSI standard. Section 5-2.1.3 of the ANSI standard requires that brakes and clutch system parts, linings, pawls, and ratchets be periodically inspected for excessive wear. Section 5-2.3.3 requires that any unsafe conditions disclosed by the frequent and periodic inspection requirements be corrected before operation of the crane is resumed.

ANSI B30.5-1968, Safety Code for Crawler, Locomotive and Truck Cranes, requires that crawler, locomotive, and truck cranes be operated only by designated operators, learners under the direct supervision of designated operators, maintenance and test personnel when necessary in the performance of their duties, and inspectors. Under ANSI B30.5-1968, a designated operator is one selected or assigned by the employer or the employer's representative as being qualified to operate the cranes. A qualified person is also defined at 1926.32(m) as one who, by professional standing, or who by extensive knowledge, training, and experience, has successfully demonstrated his ability to solve or resolve problems relating to the subject matter, the work, or the project.

The practice of leaving a mobile crane with a suspended load over a weekend is prohibited by section 1926.550(b)(2) due to the statement: "All crawler, truck, or locomotive cranes in use shall meet the applicable requirements for operation as prescribed in the ANSI B30.5-1968, Safety Code for Crawler, Locomotive and Truck Cranes." As stipulated in the Operating Practices found in Section 5-3.1.3 of ANSI B30.5-1968, under paragraph (f): "Before leaving his crane unattended, the operator shall: (1) Land any attached load, bucket, lifting magnet, or other device." Since the OSHA regulation adopts ANSI B30.5-1968, the provisions contained therein, as mentioned above, have the force of law, and compliance with them is mandatory.

(c) Hammerhead Tower Cranes

(c)(1)

Adequate clearance shall be maintained between moving and rotating structures of the crane and fixed objects to allow the passage of employees without harm.

(c)(2)

Each employee required to perform duties on the horizontal boom of hammerhead tower cranes shall be protected against falling by guardrails or by a personal fall arrest system in conformance with subpart M {Fall Protection} of this part.

(c)(3)

Buffers shall be provided at both ends of travel of the trolley.

(c)(4)

Cranes mounted on rail tracks shall be equipped with limit switches limiting the travel of the crane on the track and stops or buffers at each end of the tracks.

(c)(5)

All hammerhead tower cranes in use shall meet the applicable requirements for design, construction, installation, testing, maintenance, inspection, and operation as prescribed by the manufacturer.

(d) Overhead and Gantry Cranes

(d)(1)

The rated load of the crane shall be plainly marked on each side of the crane, and if the crane has more than one hoisting unit, each hoist shall have its rated load marked on it or its load block, and this marking shall be clearly legible from the ground or floor.

(d)(2)

Bridge trucks shall be equipped with sweeps which extend below the top of the rail and project in front of the truck wheels.

(d)(3)

Except for floor-operated cranes, a gong or other effective audible warning signal shall be provided for each crane equipped with a power traveling mechanism.

(d)(4)

All overhead and gantry cranes in use shall meet the applicable requirements for design, construction, installation, testing, maintenance, inspection, and operation as prescribed in the ANSI B30.2.0-1967, Safety Code for Overhead and Gantry Cranes.

(e) Derricks

All derricks in use shall meet the applicable requirements for design, construction, installation, inspection, testing, maintenance, and operation as prescribed in American National Standards Institute B30.6-1969, Safety Code for Derricks.

(f) Floating Cranes and Derricks

(f)(1)

Mobile cranes mounted on barges.

(f)(1)(i)

When a mobile crane is mounted on a barge, the rated load of the crane shall not exceed the original capacity specified by the manufacturer.

(f)(1)(ii)

A load rating chart, with clearly legible letters and figures, shall be provided with each crane, and securely fixed at a location easily visible to the operator.

(f)(1)(iii)

When load ratings are reduced to stay within the limits for list of the barge with a crane mounted on it, a new load rating chart shall be provided.

(f)(1)(iv)

Mobile cranes on barges shall be positively secured.

(f)(2)

Permanently mounted floating cranes and derricks.

(f)(2)(i)

When cranes and derricks are permanently installed on a barge, the capacity and limitations of use shall be based on competent design criteria.

(f)(2)(ii)

A load rating chart with clearly legible letters and figures shall be provided and securely fixed at a location easily visible to the operator.

(f)(2)(iii)

Floating cranes and floating derricks in use shall meet the applicable requirements for design, construction, installation, testing, maintenance, and operation as prescribed by the manufacturer.

(f)(3)

Protection of employees working on barges. The employer shall comply with the applicable requirements for protection of employees working onboard marine vessels specified in 1926.605.

(g) Crane or Derrick Suspended Personnel Platforms

> *On August 2, 1988, 29CFR 1926.550 was modified by the addition of a paragraph (g), which codifies OSHA Instruction STD 1-11.2B. 53 Fed. Reg. 29116 (1988). Section 1910.180(h)(3)(v) has not been modified in this fashion. The Instruction provides that employers will not be cited for violating section 1926.550(b)(2) by using cranes "to hoist and suspend employees on a work platform...when such action results in the least hazardous exposure to employees" – if a number of criteria are met. The criteria most important to this case require that a full-cycle operational test lift shall be made prior to lifting of employees. The platform shall carry twice the intended load during the test lift.*

(g)(1)

Scope, application and definitions -

(g)(1)(i)

Scope and application. This standard applies to the design, construction, testing, use and maintenance of personnel platforms, and the hoisting of personnel platforms on the load lines of cranes or derricks.

(g)(1)(ii) Definitions

Definitions. For the purposes of this paragraph (g), the following definitions apply:

(g)(1)(ii)(A)

"Failure" means load refusal, breakage, or separation of components.

(g)(1)(ii)(B)

"Hoist" (or hoisting) means all crane or derrick functions such as lowering, lifting, swinging, booming in and out or up and down, or suspending a personnel platform.

(g)(1)(ii)(C)

"Load refusal" means the point where the ultimate strength is exceeded.

(g)(1)(ii)(D)

"Maximum intended load" means the total load of all employees, tools, materials, and other loads reasonably anticipated to be applies to a personnel platform or personnel platform component at any one time.

(g)(1)(ii)(E)

"Runway" means a firm, level surface designed, prepared and designated as a path of travel for the weight and configuration of the crane being used to lift and travel with the crane suspended platform. An existing surface may be used as long as it meets these criteria.

(g)(2)

General requirements. The use of a crane or derrick to hoist employees on a personnel platform is prohibited, except when the erection, use, and dismantling of conventional means of reaching the worksite, such as a personnel hoist, ladder, stairway, aerial lift, elevating work platform or scaffold, would be more hazardous or is not possible because of structural design or worksite conditions.

The final rule regarding crane- or derrick-suspended personnel platforms was published August 2, 1988, with an original effective date of October 3, 1988. Unfortunately, anti-two-blocking devices were not as readily available from manufacturers as had been anticipated, and it was not OSHA's intent to enforce a standard that could not be reasonably complied with. Therefore, on June 26, 1989, OSHA issued interim enforcement guidance to the field and delayed the effective compliance date to December 31, 1989, to allow employers additional time for delivery and installation of anti-two-blocking devices.

Information Date: 08/02/1988
Federal Register #: 53:29116-29141
Standard Number: 1926.550
Type: Final
Agency: OSHA
Subject: Crane or Derrick Suspended Personnel Platforms.
CFR Title: 29

Abstract: OSHA hereby amends its Construction Standards for Cranes and Derricks [1926.550], by adding a new paragraph (g) to prohibit the use of cranes or derricks to hoist personnel except in the situation where no safe alternative is possible, and as long as the requirements for such hoisting set out in paragraph (g) are satisfied.

Effective date: December 31, 1989.

The preamble to the final rule explains that the record supports OSHA's determination that hoisting a suspended personnel platform with a crane or derrick constitutes a significant hazard to hoisted employees and that it is prohibited unless there is no feasible safe alternative. The preamble further states that cranes and derricks are not manufactured for use as personnel hoists, thus supporting the claim that some cranes cannot be retrofitted with two-block devices. As indicated by the Summary of the Regulatory impact Analysis (53 FR 29139), OSHA is aware that "some cranes, especially older mechanical ones, would require considerable modification in order to comply with the standard." The Agency has anticipated that some of those older cranes would be too difficult or expensive to retrofit.

Recognizing this fact and the general prohibition against hoisting personnel with cranes or derricks, it would be appropriate for OSHA to make special allowances or approve the use of any equipment or procedures that cannot comply with the current requirements of the standard. It is the responsibility of the employer, who determines that the use of a personnel platform is the only safe method available, to provide equipment that complies with all elements of the regulations. If a particular lattice boom crane cannot be adapted to use anti-two-blocking devices, it shall not be used to hoist personnel.

(g)(3) Cranes and Derricks

(g)(3)(i)

Operational criteria.

(g)(3)(i)(A)

Hoisting of the personnel platform shall be performed in a slow, controlled, cautious manner with no sudden movements of the crane or derrick, or the platform.

(g)(3)(i)(B)

Load lines shall be capable of supporting, without failure, at least seven times the maximum intended load, except that where rotation resistant rope is used, the lines shall be capable of supporting without failure, at least ten times the maximum intended load. The required design factor is achieved by taking the current safety factor of 3.5 (required under 1926.550(b)(2) and applying the 50 per cent derating of the crane capacity which is required by 1926.550(g)(3)(i)(F).

(g)(3)(i)(C)

Load and boom hoist drum brakes, swing brakes, and locking devices such as pawls or dogs shall be engaged when the occupied personnel platform is in a stationary position.

(g)(3)(i)(D)

The crane shall be uniformly level within one percent of level grade and located on firm footing. Cranes equipped with outriggers shall have them all fully deployed following manufacturer's specifications, insofar as applicable, when hoisting employees.

(g)(3)(i)(E)

The total weight of the loaded personnel platform and related rigging shall not exceed 50 percent of the rated capacity for the radius and configuration of the crane or derrick.

(g)(3)(i)(F)

The use of machines having live booms (booms in which lowering is controlled by a brake without aid from other devices which slow the lowering speeds) is prohibited.

(g)(3)(ii)

Instruments and components.

(g)(3)(ii)(A)

Cranes and derricks with variable angle booms shall be equipped with a boom angle indicator, readily visible to the operator.

(g)(3)(ii)(B)

Cranes with telescoping booms shall be equipped with a device to indicate clearly to the operator, at all times, the boom's extended length or an accurate determination of the load radius to be used during the lift shall be made prior to hoisting personnel.

(g)(3)(ii)(C)

A positive acting device shall be used which prevents contact between the load block or overhaul ball and the boom tip (anti-two-blocking device), or a system shall be used which deactivates the hoisting action before damage occurs in the event of a two-blocking situation (two-block damage prevention feature).

(g)(3)(ii)(D)

The load line hoist drum shall have a system or device on the power train, other than the load hoist brake, which regulates the lowering rate of speed of the hoist mechanism (controlled load lowering.) Free fall is prohibited.

(g)(4)

Personnel Platforms.

> *Personnel platforms that are suspended from the load line and used in construction are covered by section 29 CFR 1926.550(g). Under this standard there is no requirement for controls at the platform station. In addition, there is no specific provision for suspended personnel platforms in Part 1910. The governing provision,*

therefore, is general provision 1910.180(h)(3)(v), which prohibits hoisting, lowering, swinging, or traveling while anyone is on the load or hook. OSHA has determined, however, that when the use of a conventional means of access to an elevated worksite would be impossible or more hazardous, a violation of 1910.180(h)(3)(v) will be treated as de minimis if the employer has complied with the provisions set forth in 1926.550(g)(3), (4), (5), (6), (7), and (8).

Any use of a crane to hoist employees on a personnel platform is prohibited, except when the erection, use, and dismantling of conventional means of reaching the worksite, such as a personnel hoist, ladder, stairway, aerial lift, elevating work platform, or scaffold would be more hazardous, or is not possible because of structural design or worksite conditions.

Employee safety, rather than practicality or convenience, must be the basis for the use of a crane or derrick to lift personnel. Consequently, the use of crane- or derrick-suspended personnel platforms is prohibited by OSHA except under very limited circumstances. When determining whether the use of a crane- or derrick-suspended personnel platforms is justified, all relevant factors must be considered. For example, exposure to cardiovascular stress is only one of many factors to be assessed when making this decision. Another factor is whether or not other personnel are available to do the work. However, assuming, for the sake of argument, that a particular employee has been thoroughly examined by a physician and a determination made by the physician that the employee should use a crane- or derrick-suspended personnel platform because of cardiovascular risk, all other employees would still be required to use conventional means of access, such as stairs, thus minimizing the exposure to the hazards of using a crane- or derrick-suspended personnel platform.

In order to make a determination of the risk of exposure to cardiovascular stress, the physician must examine each individual employee. Only those found to be at risk would be allowed to use a crane or derrick suspended personnel platform (assuming, as stated above, those individuals are needed to perform the task). The physician must be given complete and detailed information on the anticipated stress, including, but not limited to, the number of steps, rate of ascent, frequency of ascent, weight of equipment, and any other factors the physician may deem necessary for a complete evaluation.

Even if the examining physician determines that a particular employee would be subject to cardiovascular stress, the employer must consider whether reasonable changes can be made that will avoid the risk. For example, equipment can be hoisted rather than carried. Pre-planning and modification of work schedules can possibly reduce the number of ascents that an employee must make.

The following documents should be reviewed by those needing additional platform design and construction information: ASME B30.5-3.2.2, Personnel Lifting; ANSI 10.28, Safety Requirements for Work Platforms Suspended From Cranes or Derricks for Construction and Demolition Operation.

(g)(4)(i)

Design criteria.

(g)(4)(i)(A)

The personnel platform and suspension system shall be designed by a qualified engineer or a qualified person competent in structural design.

(g)(4)(i)(B)

The suspension system shall be designed to minimize tipping of the platform due to movement of employees occupying the platform.

(g)(4)(i)(C)

The personnel platform itself, except the guardrail system and personnel fall arrest system anchorages, shall be capable of supporting, without failure, its own load. Criteria for guardrail systems and personal fall arrest system anchorages are contained in subpart M of this Part.

(g)(4)(ii)

Platform specifications.

(g)(4)(ii)(A)

Each personnel platform shall be equipped with a guardrail system which meets the requirements of Subpart M, and shall be enclosed at least from the toeboard to mid-rail with either solid construction or expanded metal having openings no greater than 1/2 inch (1.27 cm).

(g)(4)(ii)(B)

A grab rail shall be installed inside the entire perimeter of the personnel platform.

(g)(4)(ii)(C)

Access gates, if installed, shall not swing outward during hoisting.

(g)(4)(ii)(D)

Access gates, including sliding or folding gates, shall be equipped with a restraining device to prevent accidental opening.

(g)(4)(ii)(E)

Headroom shall be provided which allows employees to stand upright in the platform.

(g)(4)(ii)(F)

In addition to the use of hard hats, employees shall be protected by overhead protection on the personnel platform when employees are exposed to falling objects.

(g)(4)(ii)(G)

All rough edges exposed to contact by employees shall be surfaced or smoothed in order to prevent injury to employees from punctures or lacerations.

(g)(4)(ii)(H)

All welding of the personnel platform and its components shall be performed by a qualified welder familiar with the weld grades, types and material specified in the platform design.

(g)(4)(ii)(I)

The personnel platform shall be conspicuously posted with a plate or other permanent marking which indicates the weight of the platform, and its rated load capacity or maximum intended load.

(g)(4)(iii)

Personnel platform loading.

(g)(4)(iii)(A)

The personnel platform shall not be loaded in excess of its rated load capacity, When a personnel platform does not have a rated load capacity then the personnel platform shall not be loaded in excess of its maximum intended load.

(g)(4)(iii)(B)

The number of employees occupying the personnel platform shall not exceed the number required for the work being performed.

(g)(4)(iii)(C)

Personnel platforms shall be used only for employees, their tools and the materials necessary to do their work, and shall not be used to hoist only materials or tools when not hoisting personnel.

(g)(4)(iii)(D)

Materials and tools for use during a personnel lift shall be secured to prevent displacement.

(g)(4)(iii)(E)

Materials and tools for use during a personnel lift shall be evenly distributed within the confines of the platform while the platform is suspended.

(g)(4)(iv)

Rigging.

(g)(4)(iv)(A)

When a wire rope bridle is used to connect the personnel platform to the load line, each bridle leg shall be connected to a master link or shackle in such a manner to ensure that the load is evenly divided among the bridle legs.

Although OSHA standards do not address the attachment of personnel platforms to the crane boom, America National Standards Institute's recommended industry standard ANSI B30.5(b) - 1991, Section 5-3.2.2 addresses this practice. In instances where OSHA standards do not specifically address a particular activity or hazard, ANSI standards can be used by OSHA as a basis for a citation under 5(a)(1) of the Act.

(g)(4)(iv)(B)

Hooks on overhaul ball assemblies, lower load blocks, or other attachment assemblies shall be of a type that can be closed and locked, eliminating the hook throat opening. Alternatively, an alloy anchor type shackle with a bolt, nut and retaining pin may be used.

(g)(4)(iv)(C)

Wire rope, shackles, rings, master links, and other rigging hardware must be capable of supporting, without failure, at least five times the maximum intended load applied or transmitted to that component. Where rotation resistant rope is used, the slings shall be capable of supporting without failure at least ten times the maximum intended load.

(g)(4)(iv)(D)

All eyes in wire rope slings shall be fabricated with thimbles.

(g)(4)(iv)(E)

Bridles and associated rigging for attaching the personnel platform to the hoist line shall be used only for the platform and the necessary employees, their tools and the materials necessary to do their work and shall not be used for any other purpose when not hoisting personnel.

(g)(5)

Trial lift, inspections and proof testing.

(g)(5)(i)

A trial lift with the unoccupied personnel platform loaded at least to the anticipated lift weight shall be made from ground level, or any other location where employees will enter the platform to each location at which the personnel platform is to be hoisted and positioned. This trial lift shall be performed immediately prior to placing personnel on the platform. The operator shall determine that all systems, controls and safety devices are activated and functioning properly; that no interferences exist; and that all configurations necessary to reach those work locations will allow the operator to remain under the 50 percent limit of the hoist's rated capacity. Materials and tools to be used during the actual lift can be loaded in the platform, as provided in paragraphs (g)(4)(iii)(D), and (E) of this section for the trial lift. A single trial lift may be performed at one time for all locations that are to be reached from a single set up position.

(g)(5)(ii)

The trial lift shall be repeated prior to hoisting employees whenever the crane or derrick is moved and set up in a new location or returned to a previously used location. Additionally , the trial lift shall be repeated when the lift route is changed unless the operator determines that the route change is not significant (i.e. the route change would not affect the safety of hoisted employees.)

(g)(5)(iii)

After the trial lift, and just prior to hoisting personnel, the platform shall be hoisted a few inches and inspected to ensure that it is secure and properly balanced. Employees shall not be hoisted unless the following conditions are determined to exist:

(g)(5)(iii)(A)

Hoist ropes shall be free of kinks;

(g)(5)(iii)(B)

Multiple part lines shall not be twisted around each other;

(g)(5)(iii)(C)

The primary attachment shall be centered over the platform, and

(g)(5)(iii)(D)

The hoisting system shall be inspected if the load rope is slack to ensure all ropes are properly stated on drums and in sheaves.

(g)(5)(iv)

A visual inspection of the crane or derrick, rigging, personnel platform, and the crane or derrick base support or ground shall be conducted by a competent person immediately after the trial lift to determine whether the testing has exposed any defect or produced any adverse effect upon any component or structure.

(g)(5)(v)

Any defects found during inspections which create a safety hazard shall be corrected before hoisting personnel.

(g)(5)(vi)

At each job site, prior to hoisting employees on the personnel platform, and after any repair or modification, the platform and rigging shall be proof tested to 125 percent of the platform's rated capacity by holding it in a suspended position for five minutes with the test load evenly distributed on the platform (this may be done concurrently with the trial lift). After proof testing, a competent person shall inspect the platform and rigging. Any deficiencies found shall be corrected and another proof test shall be conducted. Personnel hoisting shall not be conducted until the proof testing requirements are satisfied.

(g)(6)

Work practices.

(g)(6)(i)

Employees shall keep all parts of the body inside the platform during raising lowering, and positioning. This provision does not apply to an occupant of the platform performing the duties of a signal person.

(g)(6)(ii)

Before employees exit or enter a hoisted personnel platform that is not landed, the platform shall be secured to the structure where the work is to be performed, unless securing to the structure creates an unsafe situation.

(g)(6)(iii)

Tag lines shall be used unless their use creates an unsafe condition.

(g)(6)(iv)

The crane or derrick operator shall remain at the controls at all times when the crane engine is running and the platform is occupied.

(g)(6)(v)

Hoisting of employees shall be promptly discontinued upon indication of any dangerous weather conditions or other impending danger.

(g)(6)(vi)

Employees being hoisted shall remain in continuous sight of and in direct communication with the operator or signal person. In those situations where direct visual contact with the operator is not possible, and the use of a signal person would create a greater hazard for the person, direct communication alone such as by radio may be used.

(g)(6)(vii)

Except over water, employees occupying the personnel platform shall use a body belt/harness system with lanyard appropriately attached to the lower load block or overhaul ball, or to a structural member within the personnel platform capable of supporting a fall impact for employees using the anchorage. When working over water, the requirements of 1926.106 shall apply.

(g)(6)(viii)

No lifts shall be made on another of the crane's or derrick's loadlines while personnel are suspended on a platform.

> *Hoisting employees on the secondary line of a telescoping crane while a load is hoisted on the main load line is prohibited by section (g)(6)(viii). OSHA stated clearly in the preamble to the final rule for Crane or Derrick Suspended Personnel Platforms (53 FR 29136, August 2, 1988) that cranes may not be used to hoist personnel while other loads are being hoisted. The use of other lines would endanger the hoisted personnel due to the risk of entanglement.*

Compliance with section 1926.550(g)(6)(viii) could be achieved by using two cranes, one to lift and hold the load and the other to lift and position the personnel in the platform. However, if the circumstances at a particular worksite will not permit the use of two cranes and alternative procedures using a single crane could be implemented that would provide equivalent protection for affected employees, an application for a permanent variance would be appropriate. Information on variance procedures can be obtained from the Office of Variance Determination in the Directorate of Technical Support at (202) 219-7193.

(g)(7) Traveling

(g)(7)(i)

Hoisting of employees while the crane is traveling is prohibited, except for portal, tower and locomotive cranes, or where the employer demonstrates that there is no less hazardous way to perform the work.

(g)(7)(ii)

Under any circumstances where a crane would travel while hoisting personnel, the employer shall implement the following procedures to safeguard employees:

(g)(7)(ii)(A)

Crane travel shall be restricted to a fixed track or runway;

(g)(7)(ii)(B)

Travel shall be limited to the load radius of the boom used during the lift; and

(g)(7)(ii)(C)

The boom must be parallel to the direction of travel.

(g)(7)(ii)(D)

A complete trial run shall be performed to test the route of travel before employees are allowed to occupy the platform. This trial run can be performed at the same time as the trial lift required by paragraph (g)(5)(i) of this section which tests the route of the lift.

(g)(7)(ii)(E)

If travel is done with a rubber tired-carrier, the condition and air pressure of the tires shall be checked. The chart capacity for lifts on rubber shall be used for application of the 50 percent reduction of rated capacity. Notwithstanding paragraph (g)(5)(i)(E) of this section, outriggers may be partially retracted as necessary for travel.

(g)(8)

Pre-lift meeting.

(g)(8)(i)

A meeting attended by the crane or derrick operator, signal person(s)(if necessary for the lift), employee(s) to be lifted, and the person responsible for the task to be performed shall be held to review the appropriate requirements of paragraph (g) of this section and the procedures to be followed.

(g)(8)(ii)

This meeting shall be held prior to the trial lift at each new work location, and shall be repeated for any employees newly assigned to the operation.

> *Note: For additional information see Appendix J, Crane or Derrick Suspended Platforms, OSHA 3100, 1993 (Revised)*

[44 FR 8577, Feb. 9, 1979; 44 FR 20940, Apr. 6, 1979, as amended at 52 FR 36382, Sept. 28, 1987; 53 FR 29139, Aug. 2, 1988; 54 FR 15406, Apr. 18, 1989; 54 FR 24334, June 7, 1989; 58 FR 35183, June 30, 1993, 59 FR 40730, Aug. 9, 1994; 61 FR 5507, Feb. 13, 1996]

4.E. OSHA 1926.251 — Rigging Equipment for Material Handling

Standard Number: 1926.251

Standard Title: Rigging equipment for material handling.

SubPart Number: H

SubPart Title: Materials Handling, Storage, Use, and Disposal

Applicable Standard: 1910.184(a)

Applicable Standard: 1910.184(c)(2)

Applicable Standard: 1910.184(c)(3)

Applicable Standard: 1910.184(c)(5)

Applicable Standard: 1910.184(c)(7)

Applicable Standard: 1910.184(c)(10)

(a) General

(a)(1)

Rigging equipment for material handling shall be inspected prior to use on each shift and as necessary during its use to ensure that it is safe. Defective rigging equipment shall be removed from service.

(a)(2)

Rigging equipment shall not be loaded in excess of its recommended safe working load, as prescribed in Tables H-1 through H-20 in this subpart, following 1926.252(e) for the specific equipment.

> *The tables in this standard (1926.251) that are identical to those in 1910.184 are identified as such and are not repeated here. Instead, the page number is given where the corresponding table can be found in this text.*

(a)(3)

Rigging equipment, when not in use, shall be removed from the immediate work area so as not to present a hazard to employees.

(a)(4)

Special custom design grabs, hooks, clamps, or other lifting accessories, for such units as modular panels, prefabricated structures and similar materials, shall be marked to indicate the safe working loads and shall be proof-tested prior to use to 125 percent of their rated load.

(a)(5)

"Scope." This section applies to slings used in conjunction with other material handling equipment for the movement of material by hoisting, in employments covered by this part. The types of slings covered are those made from alloy steel chain, wire rope, metal mesh, natural or synthetic fiber rope (conventional three strand construction), and synthetic web (nylon, polyester, and polypropylene).

(a)(6)

"Inspections." Each day before being used, the sling and all fastenings and attachments shall be inspected for damage or defects by a competent person designated by the employer. Additional inspections shall be performed during sling use, where service conditions warrant. Damaged or defective slings shall be immediately removed from service.

(b) Alloy Steel Chains

(b)(1)

Welded alloy steel chain slings shall have permanently affixed durable identification stating size, grade, rated capacity, and sling manufacturer.

(b)(2)

Hooks, rings, oblong links, pear-shaped links, welded or mechanical coupling links, or other attachments, when used with alloy steel chains, shall have a rated capacity at least equal to that of the chain.

(b)(3)

Job or shop hooks and links, or makeshift fasteners, formed from bolts, rods, etc., or other such attachments, shall not be used.

(b)(4)

Rated capacity (working load limit) for alloy steel chain slings shall conform to the values shown in Table H-1.

(b)(5)

Whenever wear at any point of any chain link exceeds that shown in Table H-2, the assembly shall be removed from service.

(b)(6)

"Inspections."

(b)(6)(i)

In addition to the inspection required by other paragraphs of this section, a thorough periodic inspection of alloy steel chain slings in use shall be made on a regular basis, to be determined on the basis of (A) frequency of sling use; (B) severity of service conditions; (C) nature of lifts being made; and (D) experience gained on the service life of slings used in similar circumstances. Such inspections shall in no event be at intervals greater than once every 12 months.

(b)(6)(ii)

The employer shall make and maintain a record of the most recent month in which each alloy steel chain sling was thoroughly inspected, and shall make such record available for examination.

(c) Wire Rope

(c)(1)

Tables H-3 through H-14 shall be used to determine the safe working loads of various sizes and classifications of improved plow steel wire rope and wire rope slings with various types of terminals. For sizes, classifications, and grades not included in these tables, the safe working load recommended by the manufacturer for specific, identifiable products shall be followed, provided that a safety factor of not less than 5 is maintained.

(c)(2)

Protruding ends of strands in splices on slings and bridles shall be covered or blunted.

(c)(3)

Wire rope shall not be secured by knots, except on haul back lines on scrapers.

(c)(4)

The following limitations shall apply to the use of wire rope:

(c)(4)(i)

An eye splice made in any wire rope shall have not less than three full tucks. However, this requirement shall not operate to preclude the use of another form of splice or connection which can be shown to be as efficient and which is not otherwise prohibited.

(c)(4)(ii)

Except for eye splices in the ends of wires and for endless rope slings, each wire rope used in hoisting or lowering, or in pulling loads, shall consist of one continuous piece without knot or splice.

(c)(4)(iii)

Eyes in wire rope bridles, slings, or bull wires shall not be formed by wire rope clips or knots.

(c)(4)(iv)

Wire rope shall not be used if, in any length of eight diameters, the total number of visible broken wires exceeds 10 percent of the total number of wires, or if the rope shows other signs of excessive wear, corrosion, or defect.

(c)(5)

When U-bolt wire rope clips are used to form eyes, Table H-20 shall be used to determine the number and spacing of clips.

(c)(5)(i)

When used for eye splices, the U-bolt shall be applied so that the "U" section is in contact with the dead end of the rope.

(c)(6)

Slings shall not be shortened with knots or bolts or other makeshift devices.

(c)(7)

Sling legs shall not be kinked.

(c)(8)

Slings used in a basket hitch shall have the loads balanced to prevent slippage.

(c)(9)

Slings shall be padded or protected from the sharp edges of their loads.

(c)(10)

Hands or fingers shall not be placed between the sling and its load while the sling is being tightened around the load.

(c)(11)

Shock loading is prohibited.

(c)(12)

A sling shall not be pulled from under a load when the load is resting on the sling.

(c)(13)

"Minimum sling lengths."

(c)(13)(i)

Cable laid and 6 X 19 and 6 X 37 slings shall have minimum clear length of wire rope 10 times the component rope diameter between splices, sleeves or end fittings.

(c)(13)(ii)

Braided slings shall have a minimum clear length of wire rope 40 times the component rope diameter between the loops or end fittings.

(c)(13)(iii)

Cable laid grommets, strand laid grommets and endless slings shall have a minimum circumferential length of 96 times their body diameter.

(c)(14)

"Safe operating temperatures." Fiber core wire rope slings of all grades shall be permanently removed from service if they are exposed to temperatures in excess of 200 deg. F (93.33 deg. C). When non-fiber core wire rope slings of any grade are used at temperatures above 400 deg. F (204.44 deg. C) or below minus 60 deg. F (15.55 deg. C), recommendations of the sling manufacturer regarding use at that temperature shall be followed.

(c)(15)

"End attachments."

(c)(15)(i)

Welding of end attachments, except covers to thimbles, shall be performed prior to the assembly of the sling.

(c)(15)(ii)

All welded end attachments shall not be used unless proof tested by the manufacturer or equivalent entity at twice their rated capacity prior to initial use. The employer shall retain a certificate of proof test, and make it available for examination.

(d) Natural Rope and Synthetic Fiber

(d)(1)

General. When using natural or synthetic fiber rope slings, Tables H-15, 16, 17, and 18 shall apply.

(d)(2)

All splices in rope slings provided by the employer shall be made in accordance with fiber rope manufacturers recommendations.

(d)(2)(i)

In manila rope, eye splices shall contain at least three full tucks, and short splices shall contain at least six full tucks (three on each side of the center line of the splice).

(d)(2)(ii)

In layered synthetic fiber rope, eye splices shall contain at least four full tucks, and short splices shall contain at least eight full tucks (four on each side of the center line of the splice).

(d)(2)(iii)

Strand end tails shall not be trimmed short (flush with the surface of the rope) immediately adjacent to the full tucks. This precaution applies to both eye and short splices and all types of fiber rope. For fiber ropes under 1-inch diameter, the tails shall project at least six rope diameters beyond the last full tuck. For fiber ropes 1-inch diameter and larger, the tails shall project at least 6 inches beyond the last full tuck. In applications where the projecting tails may be objectionable, the tails shall be tapered and spliced into the body of the rope using at least two additional tucks (which will require a tail length of approximately six rope diameters beyond the last full tuck).

(d)(2)(iv)

For all eye splices, the eye shall be sufficiently large to provide an included angle of not greater than 60 deg. at the splice when the eye is placed over the load or support.

(d)(2)(v)

Knots shall not be used in lieu of splices.

(d)(3)

"Safe operating temperatures." Natural and synthetic fiber rope slings, except for wet frozen slings, may be used in a temperature range from minus 20 deg. F (-28.88 deg. C) to plus 180 deg. F (82.2 deg. C) without decreasing the working load limit. For operations outside this temperature range and for wet frozen slings, the sling manufacturer's recommendations shall be followed.

(d)(4)

"Splicing." Spliced fiber rope slings shall not be used unless they have been spliced in accordance with the following minimum requirements and in accordance with any additional recommendations of the manufacturer:

(d)(4)(i)

In manila rope, eye splices shall consist of at least three full tucks, and short splices shall consist of at least six full tucks, three on each side of the splice center line.

(d)(4)(ii)

In synthetic fiber rope, eye splices shall consist of at least four full tucks, and short splices shall consist of at least eight full tucks, four on each side of the center line.

(d)(4)(iii)

Strand end tails shall not be trimmed flush with the surface of the rope immediately adjacent to the full tucks. This applies to all types of fiber rope and both eye and short splices. For fiber rope under 1 inch (2.54 cm) in diameter, the tail shall project at least six rope diameters beyond the last full tuck. For fiber rope 1 inch (2.54 cm) in diameter and larger, the tail shall project at least 6 inches (15.24 cm) beyond the last full tuck. Where a projecting tail interferes with the use of the sling, the tail shall be tapered and spliced into the body of the rope using at lest two additional tucks (which will require a tail length of approximately six rope diameters beyond the last full tuck).

(d)(4)(iv)

Fiber rope slings shall have a minimum clear length of rope between eye splices equal to 10 times the rope diameter.

(d)(4)(v)

Knots shall not be used in lieu of splices.

(d)(4)(vi)

Clamps not designed specifically for fiber ropes shall not be used for splicing.

(d)(4)(vii)

For all eye splices, the eye shall be of such size to provide an included angle of not greater than 60 degrees at the splice when the eye is placed over the load or support.

(d)(5)

"End attachments." Fiber rope slings shall not be used if end attachments in contact with the rope have sharp edges or projections.

(d)(6)

"Removal from service." Natural and synthetic fiber rope slings shall be immediately removed from service if any of the following conditions are present:

(d)(6)(i)

Abnormal wear.

(d)(6)(ii)

Powdered fiber between strands.

(d)(6)(iii)

Broken or cut fibers.

(d)(6)(iv)

Variations in the size or roundness of strands.

(d)(6)(v)

Discoloration or rotting.

(d)(6)(vi)

Distortion of hardware in the sling.

(e) Synthetic Webbing (nylon, polyester, and polypropylene)

(e)(1)

The employer shall have each synthetic web sling marked or coded to show:

(e)(1)(i)

Name or trademark of manufacturer.

(e)(1)(ii)

Rated capacities for the type of hitch.

(e)(1)(iii)

Type of material.

(e)(2)

Rated capacity shall not be exceeded.

(e)(3)

"Webbing." Synthetic webbing shall be of uniform thickness and width and selvage edges shall not be split from the webbing's width.

(e)(4)

"Fittings." Fittings shall be:

(e)(4)(i)

Of a minimum breaking strength equal to that of the sling; and

(e)(4)(ii)

Free of all sharp edges that could in any way damage the webbing.

(e)(5)

"Attachment of end fittings to webbing and formation of eyes." Stitching shall be the only method used to attach end fittings to webbing and to form eyes. The thread shall be in an even pattern and contain a sufficient number of stitches to develop the full breaking strength of the sling.

(e)(6)

"Environmental conditions." When synthetic web slings are used, the following precautions shall be taken:

(e)(6)(i)

Nylon web slings shall not be used where fumes, vapors, sprays, mists or liquids of acids or phenolics are present.

(e)(6)(ii)

Polyester and polypropylene web slings shall not be used where fumes, vapors, sprays, mists or liquids of caustics are present.

(e)(6)(iii)

Web slings with aluminum fittings shall not be used where fumes, vapors, sprays, mists or liquids of caustics are present.

(e)(7)

"Safe operating temperatures." Synthetic web slings of polyester and nylon shall not be used at temperatures in excess of 180 deg. F (82.2 deg. C). Polypropylene web slings shall not be used at temperatures in excess of 200 deg. F (93.33 deg. C).

(e)(8)

"Removal from service." Synthetic web slings shall be immediately removed from service if any of the following conditions are present:

(e)(8)(i)

Acid or caustic burns;

(e)(8)(ii)

Melting or charring of any part of the sling surface;

(e)(8)(iii)

Snags, punctures, tears or cuts;

(e)(8)(iv)

Broken or worn stitches; or

(e)(8)(v)

Distortion of fittings.

(f) Shackles and Hooks

(f)(1)

Table H-19 shall be used to determine the safe working loads of various sizes of shack-

Table 4.E.1 Minimum Allowable Wear at any Point of link	
Chain size, inches	Minimum allowable chain size, inches
1/4	3/64
3/8	5/64
1/2	7/64
5/8	9/64
3/4	5/32
7/8	11/64
1	3/16
1 1/8	7/32
1 1/4	1/4
1 3/8	9/32
1 1/2	5/16
1 3/4	11/32

les, except that higher safe working loads are permissible when recommended by the manufacturer for specific, identifiable products, provided that a safety factor of not less than 5 is maintained.

(f)(2)

The manufacturer's recommendations shall be followed in determining the safe working loads of the various sizes and types of specific and identifiable hooks. All hooks for which no applicable manufacturer's recommendations are available shall be tested to twice the intended safe working load before they are initially put into use. The employer shall maintain a record of the dates and results of such tests.

> *Note: The tables in this standard (1926.251) that are identical to those in 1910.184 are identified as such and are not repeated here. Instead, the page number is given where the corresponding table can be found in this text.*

Table H-1. - "Rated Capacity (Working Load Limit), for Alloy Steel Chain Slings" is identical to Table N-184-1 located on page 206.

Table H-3. - "Rated Capacities for Single Leg Slings - 6x19 and 6x37 Classification Improved Plow Steel Grade Rope With Fiber Core (FC)" is identical to Table N-184-3 located on page 208.

Table H-4. - "Rated Capacities for Single Leg Slings - 6x19 and 6x37 Classification Improved Plow Steel Grade Rope With Independent Wire Rope Core (IWRC)" is identical to Table N-184-4 located on page 209.

Table H-5. - "Rated Capacities for Single Leg Slings - Cable Laid Rope – Mechanical Splice Only" is identical to Table N-184-5 located on page 210.

Table H-6. - "Rated Capacities for Single Leg Slings - 8-Part and 6-Part Braided Rope" is identical to Table N-184-6 located on page 211.

Table H-7. - "Rated Capacities for 2-Leg and 3-Leg Bridle Slings - 6x19 and 6x37 Classification Improved Plow Steel Grade Rope With Fiber Core (FC)" is identical to Table N-184-7 located on page 212.

Table H-8. - "Rated Capacities for 2-Leg and 3-Leg Bridle Slings - 6x19 and 6x37 Classification Improved Plow Steel Grade Rope With Independent Wire Rope Core (IWRC)" is identical to table N-184-8 located on page 213.

Table H-9. - "Rated Capacities for 2-Leg and 3-Leg Bridle Slings - Cable Laid Rope - Mechanical Splice Only 7x7x7 and 7x7x19 Construction Galvanized Aircraft Grade Rope 7x6x19 IWRC Construction Improved Plow Steel Grade Rope" is identical to Table N-184-9 located on page 214.

Table H-10. - "Rated Capacities for 2-Leg and 3-Leg Bridle Slings - 8-Part and 6-Part Braided Rope 6x7 and 6x19 Construction Improved Plow Steel Grade Rope 7x7 Construction Galvanized Aircraft Grade Rope" is identical to Table N-184-10 located on page 215.

Table H-11. - "Rated Capacities for Strand Laid Grommet - Hand Tucked Improved Plow Steel Grade Rope" is identical to Table N-184-11 located on page 216.

Table 4.E.2 (H-19) Safe Working Loads for Shackles (in tons of 2,000 pounds)

Material Size	Pin Diameter (Inches)	Safe Working Load
1/2	5/8	1.4
5/8	3/4	2.2
3/4	7/8	3.2
7/8	1	4.3
1	1 1/8	5.6
1 1/8	1 1/4	6.7
1 1/4	1 3/8	8.2
1 3/8	1 1/2	10.0
1 1/2	1 5/8	11.9
1 3/4	2	16.2
2	2 1/4	21.2

Table 4.E.3. (H-20) Number and Spacing of U-Bolt Wire Rope Clips

Improved plow steel, rope diameter (inches)	Number of clips		Minimum spacing (inches)
	Drop forged	Other Material	
1/2	3	4	3
5/8	3	4	3 3/4
3/4	4	5	4 1/2
7/8	4	5	5 1/4
1	5	6	6
1 1/8	6	6	6 3/4
1 1/4	6	7	7 1/2
1 3/8	7	7	8 1/4
1 1/2	7	8	9

[44 FR 8577, Feb. 9, 1979; 44 FR 20940, Apr. 6, 1979, as amended at 58 FR 35173; June 30, 1993]

Table H-12. - "Rated Capacities for Cable Laid Grommet – Hand Tucked 7x6x7 and 7x6x19 Constructions Improved Plow Steel Grade Rope 7x7x7 Construction Galvanized Aircraft Grade Rope" is identical to Table N-184-12 located on page 216.

Table H-13. - "Rated Capacities for Strand Laid Endless Slings- Mechanical Joint Improved Plow Steel Grade Rope" is identical to Table N-184-13 located on page 217.

Table H-14. - "Rated Capacities for Cable Laid Endless Slings – Mechanical Joint 7x7x7 and 7x7x19 Constructions Galvanized Aircraft Grade Rope 7x6x19 Construction Improved Plow Steel Grade Rope" is identical to Table N-184-14 located on page 217.

Table H-15. - "Rated Capacities" is identical to Table N-184-15 located on page 222.

Table H-16. - "Manila Rope Slings" is identical to Table N-184-16 located on page 224.

Table H-17. - "Nylon Rope Slings" is identical to Table N-184-17 located on page 225.

Table H-18. - "Polyester Rope Slings" is identical to Table N-184-18 located on page 226.

Table H-19. - "Polypropylene Rope Slings" is identical to Table N-184-19 located on page 227.

WIRE ROPE CLIPS

The proper performance of forged clips depends on proper manufacturing practices that include good forging techniques and accurate machining. Forged clips provide a greater rope bearing surface and more consistent strength than malleable cast iron clips. Fist grip clips provide a saddle for both the "live" and the "dead" end. Fewer forged clips are required for each termination than with malleable cast iron clips. Forged clips reduce the possibility of hidden defects that are sometimes present in malleable cast iron clips. Malleable cast iron clips should be used in only non-critical applications. ANSI, OSHA, and ASTM recommend only forged clips for critical applications.

The proper application of forged clips requires that the correct type, size, number, and installation instructions be used. Availability of a full range of sizes of forged U-bolt clips and forged fist grip clips is essential for design flexibility. The clip's size, manufacturer's logo, and a traceability code should be clearly embossed in the forging of the clip. These three elements are essential in developing total confidence in the product.

Detailed application information is needed in the proper installation of wire rope clips. This information is most effective when provided at the point of application, as well as in supporting brochures and engineering information. The manufacturer must provide this specific information because generic information will not provide all the needed application instructions. A formal application and warning system that attracts the attention of the user clearly informs the user of the factors involved in the task, and informs the user with the proper application procedures as needed. Figure 4.E.1 provides more detailed information on wire rope clips and turnbuckles.

Figure 5.E.1. Rigging Hardware – Wire Rope Clips/Turnbuckles [7]

WIRE ROPE CLIPS

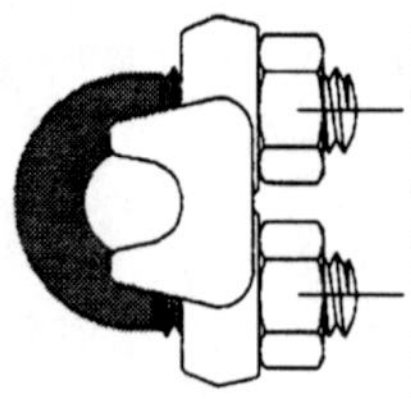

SIZE	EFFICIENCY	NUMBER OF CLIPS	TURNBACK LENGTH (IN)	TORQUE FT – lbs
1/8	80%	2	3-1/4	4.5
3/16	80%	2	3-3/4	7.5
1/4	80%	2	4-3/4	15
5/16	80%	2	5-1/4	30
3/8	80%	2	6-1/2	45
7/16	80%	2	7	65
1/2	80%	3	11-1/2	65
9/16	80%	3	12	95
5/8	80%	3	12	95
3/4	80%	4	18	130
1	90%	5	26	225

APPLY U-BOLT OVER DEAD END OF THE WIRE ROPE LIVE END OF THE ROPE RESTS IN THE SADDLE A TERMINATION IS NOT COMPLETE UNTIL IT HAS BEEN RETORQUED A SECOND TIME
NEVER SADDLE A DEAD HORSE!

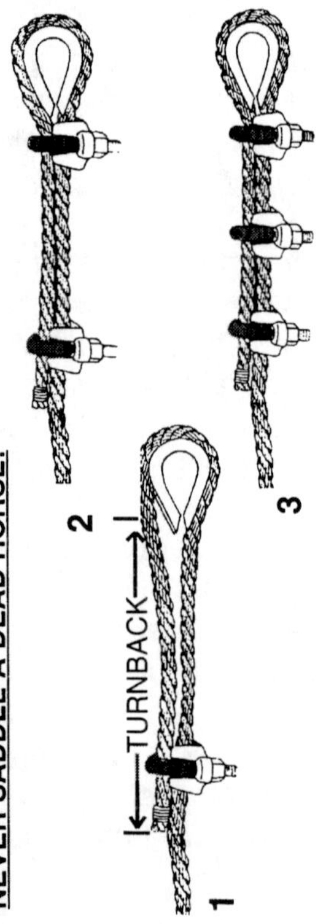

TURNBUCKLE

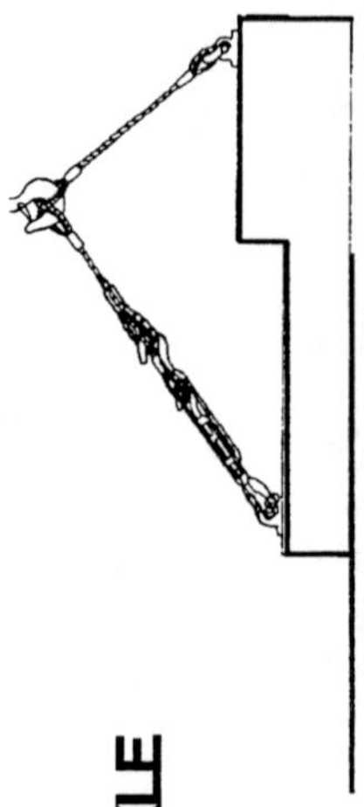

SIZE	WORKING LOAD LIMIT JAW AND EYE 5/1 DESIGN FACTOR	WORKING LOAD LIMIT HOOK END FITTING 5/1 DESIGN FACTOR
1/4	500	400
5/16	800	700
3/8	1200	1000
1/2	2200	1500
5/8	3500	2250
3/4	5200	3000
7/8	7200	4000
1	10000	5000
1-1/4	15200	5000
1-1/2	21400	7500

THE USE OF LOCKNUTS OR MOUSING IS AN EFFECTIVE METHOD OF PREVENTING TURNBUCKLES FROM ROTATING

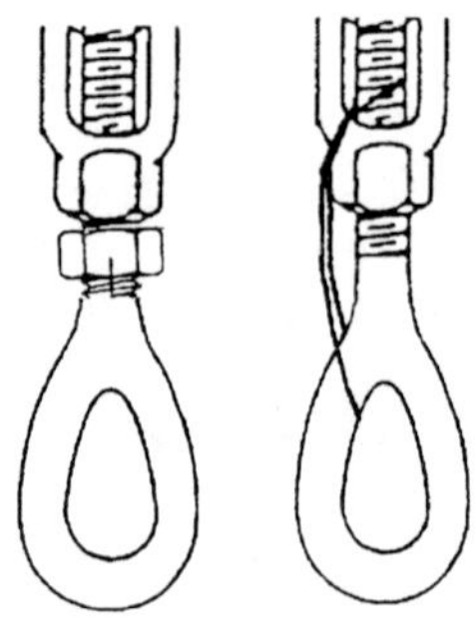

4.F. OSHA 1910.184

The booklet "Sling Safety OSHA 3072" is included in Appendix I

4.F. OSHA 1910.184 - Slings

Standard Number: 1910.184

Standard Title: Slings

Subpart Number: N

Subpart Title: Materials Handling and Storage

Produced by USDOL OSHA - Directorate of Safety Standards & Directorate of Health Standards

(a) Scope

This section applies to slings used in conjunction with other material handling equipment for the movement of material by hoisting, in employments covered by this part. The types of slings covered are those made from alloy steel chain, wire rope, metal mesh, natural or synthetic fiber rope (conventional three strand construction), and synthetic web (nylon, polyester, and polypropylene).

The alloy steel chain slings detailed in standard 1910.184 are chains made from alloy steel specified in the National Association of Chain Manufacturers (NACM) Welded Steel Chain specifications-1990 as grade 80 alloy steel chain. Any other alloy steel chain sling falling within the scope of 1910.184(a) is still governed by [1910.184(e)(5)], which states in part:

"Slings not included in this table shall be used only in accordance with the manufacturer's recommendations."

If the chain manufacturer provides an employer with information, comments, and suggestions on the terms and limitations under which their product should be used and an employer does not follow the chain manufacturer's recommendation, than that employer could be cited under the 1910.184 standard.

The sling proof test and repair certification records required in 29 CFR 1910.184 are necessary to ensure compliance with the requirements for slings in general industry. Included in this standard are the requirements for: the proof testing of all new, repaired, or reconditioned alloy steel chain slings and the certification of that proof testing; the proof testing of all wire rope slings with welded end attachments and the certification of that proof testing; and the marking, tagging, or certification of repair of metal mesh slings.

The American Society of Testing and Materials (ASTM) Specification A391 (ANSI G61.1-1968). details the composition and other details (including proof-testing) of grade 80 alloy steel chain, and the use of this ASTM specification may be inappropriate for other alloy steel chain.

Employers may use chain slings made of metal other than alloy steel in the workplace. In consideration of employee safety in the workplace, proof coil steel chain, also known as common or hardware chain, and other chain not recommended by the chain manufacturer shall not be used for hoisting purposes. Except as noted otherwise, chain slings made from other than alloy steel, designed, constructed, repaired and proof-tested for safe use, in accordance with National Consensus Standards, component manufacturer recommendations and otherwise good engineering practice of the industry, under the supervision of a qualified person, and used to support loads up to but not to exceed their rated capacity (working load limit) would be considered favorably with respect to employer compliance. A "qualified person" is defined as one with a recognized degree or professional certificate and extensive knowledge and experience in the subject field who is capable of design, analysis, and evaluation of specifications in the subject work, project or product.

(b) Definitions

"Angle of loading" is the inclination of a leg or branch of a sling measured from the horizontal or vertical plane as shown in Fig. N-184-5; provided that an angle of loading of five degrees or less from the vertical may be considered a vertical angle of loading.

"Basket hitch" is a sling configuration whereby the sling is passed under the load and has both ends, end attachments, eyes or handles on the hook or a single master link.

"Braided wire rope" is a wire rope formed by plaiting component wire ropes.

"Bridle wire rope sling" is a sling composed of multiple wire rope legs with the top ends gathered in a fitting that goes over the lifting hook.

"Cable laid endless sling-mechanical joint" is a wire rope sling made endless by joining the ends of a single length of cable laid rope with one or more metallic fittings.

"Cable laid grommet-hand tucked" is an endless wire rope sling made from one length of rope wrapped six times around a core formed by hand tucking the ends of the rope inside the six wraps.

"Cable laid rope" is a wire rope composed of six wire ropes wrapped around a fiber or wire rope core.

"Cable laid rope sling-mechanical joint" is a wire rope sling made from a cable laid rope with eyes fabricated by pressing or swaging one or more metal sleeves over the rope junction.

"Choker hitch" is a sling configuration with one end of the sling passing under the load and through an end attachment, handle or eye on the other end of the sling.

"Coating" is an elastomer or other suitable material applied to a sling or to a sling component to impart desirable properties.

"Cross rod" is a wire used to join spirals of metal mesh to form a complete fabric. (See Fig. N-184-2.)

"Designated" means selected or assigned by the employer or the employer's representative as being qualified to perform specific duties.

"Equivalent entity" is a person or organization (including an employer) which, by possession of equipment, technical knowledge and skills, can perform with equal competence the same repairs and tests as the person or organization with which it is equated.

"Fabric (metal mesh)" is the flexible portion of a metal mesh sling consisting of a series of transverse coils and cross rods.

"Female handle (choker)" is a handle with a handle eye and a slot of such dimension as to permit passage of a male handle thereby allowing the use of a metal mesh sling in a choker hitch. (See Fig. N-184-1.)

"Handle" is a terminal fitting to which metal mesh fabric is attached. (See Fig. N-184-1.)

"Handle eye" is an opening in a handle of a metal mesh sling shaped to accept a hook, shackle or other lifting device. (See Fig. N-184-1.)

"Hitch" is a sling configuration whereby the sling is fastened to an object or load, either directly to it or around it.

"Link" is a single ring of a chain.

"Male handle (triangle)" is a handle with a handle eye.

"Master coupling link" is an alloy steel welded coupling link used as an intermediate link to join alloy steel chain to master links. (See Fig. N-184-3.)

"Master link" or "gathering ring" is a forged or welded steel link used to support all members (legs) of an alloy steel chain sling or wire rope sling. (See Fig. N-184-3.)

"Mechanical coupling link" is a non-welded, mechanically closed steel link used to attach master links, hooks, etc., to alloy steel chain.

"Proof load" is the load applied in performance of a proof test.

"Proof test" is a nondestructive tension test performed by the sling manufacturer or an equivalent entity to verify construction and workmanship of a sling.

"Rated capacity" or "working load limit" is the maximum working load permitted by the provisions of this section.

"Reach" is the effective length of an alloy steel chain sling measured from the top bearing surface of the upper terminal component to the bottom bearing surface of the lower terminal component.

"Selvage edge" is the finished edge of synthetic webbing designed to prevent unraveling.

"Sling" is an assembly which connects the load to the material handling equipment.

"Sling manufacturer" is a person or organization that assembles sling components into their final form for sale to users.

"Spiral" is a single transverse coil that is the basic element from which metal mesh is fabricated. (See Fig. N-184-2.)

"Strand laid endless sling-mechanical joint" is a wire rope sling made endless from one length of rope with the ends joined by one or more metallic fittings.

"Strand laid grommet-hand tucked" is an endless wire rope sling made from one length of strand wrapped six times around a core formed by hand tucking the ends of the strand inside the six wraps.

"Strand laid rope" is a wire rope made with strands (usually six or eight) wrapped around a fiber core, wire strand core, or independent wire rope core (IWRC).

"Vertical hitch" is a method of supporting a load by a single, vertical part or leg of the sling. (See Fig. N-184-4.)

(c) Safe Operating Practices

Whenever any sling is used, the following practices shall be observed:

(c)(1)

Slings that are damaged or defective shall not be used.

(c)(2)

Slings shall not be shortened with knots or bolts or other makeshift devices.

(c)(3)

Sling legs shall not be kinked.

(c)(4)

Slings shall not be loaded in excess of their rated capacities.

(c)(5)

Slings used in a basket hitch shall have the loads balanced to prevent slippage.

(c)(6)

Slings shall be securely attached to their loads.

(c)(7)

Slings shall be padded or protected from the sharp edges of their loads.

(c)(8)

Suspended loads shall be kept clear of all obstructions.

(c)(9)

All employees shall be kept clear of loads about to be lifted and of suspended loads.

Although this standard, 1910.184(c)(9), requires that employees "be kept clear of loads about to be lifted and of suspended loads," it is not specific as to exactly when employees may place their hands upon suspended loads. Generally, performing "hands-on" work upon suspended loads (as opposed to controlling suspended loads) is prohibited unless a secondary means of support is provided to preclude unwanted movement of the load.

With regard to guiding or controlling suspended loads, tag lines should be used to the greatest practical extent, although their use is only specified in the context of structural steel erection and power transmission within OSHA's standards (29 CFR 1926.751(d) and 29 CFR 1926.953(d), respectively). However, once a load is near its final location, it may be necessary to guide the load by hand if fine load control is critical. Examples include making bolted connections during structural steel erection or aligning an equipment baseplate to pre-positioned anchor bolts. In an effort to ensure that appropriate controls are implemented, such issues could be discussed and resolved during pre-lift planning.

Given the purpose of this requirement, the word "clear" means close enough to the load that it is reasonably foreseeable that they could be hit by the load if the load should fall. While this is not a precise definition, in applying it, the results will vary depending on numerous factors, including the height of the load above the employees, the size of the load, the shape of the load, the speed at which the load is traveling, the method by which the load is fastened to the crane, the customary patterns of and physical restrictions on employee movement, and other considerations.

(c)(10)

Hands or fingers shall not be placed between the sling and its load while the sling is being tightened around the load.

(c)(11)

Shock loading is prohibited.

(c)(12)

A sling shall not be pulled from under a load when the load is resting on the sling.

(d) Inspections

Each day before being used, the sling and all fastenings and attachments shall be inspected for damage or defects by a competent person designated by the employer. Additional inspections shall be performed during sling use, where service conditions warrant. Damaged or defective slings shall be immediately removed from service.

The daily inspection of slings before use includes, as appropriate, more than just a visual glance. However, in the promulgation of this standard, OSHA did not intend for a sling to lose its utility through the inspection process. Thus, the standard requires a thorough examination of the sling, but does not require the protective seizing of nylon wrapping to be removed. Where service conditions warrant, additional inspections must be performed during sling use. Any damaged or defective sling must be immediately removed from service.

(e) Alloy Steel Chain Slings

(e)(1)

Sling identification. Alloy steel chain slings shall have permanently affixed durable identification stating size, grade, rated capacity, and reach.

(e)(2) Attachments

(e)(2)(i)

Hooks, rings, oblong links, pear shaped links, welded or mechanical coupling links or other attachments shall have a rated capacity at least equal to that of the alloy steel chain with which they are used or the sling shall not be used in excess of the rated capacity of the weakest component.

(e)(2)(ii)

Makeshift links or fasteners formed from bolts or rods, or other such attachments, shall not be used.

(e)(3) Inspections

(e)(3)(i)

In addition to the inspection required by paragraph (d) of this section, a thorough periodic inspection of alloy steel chain slings in use shall be made on a regular basis, to be determined on the basis of (A) frequency of sling use; (B) severity of service conditions; (C) nature of lifts being made; and (D) experience gained on the service life of slings used in similar circumstances. Such inspections shall in no event be at intervals greater than once every 12 months.

(e)(3)(ii)

The employer shall make and maintain a record of the most recent month in which each alloy steel chain sling was thoroughly inspected, and shall make such record available for examination.

(e)(3)(iii)

The thorough inspection of alloy steel chain slings shall be performed by a competent person designated by the employer, and shall include a thorough inspection for wear, defective welds, deformation and increase in length. Where such defects or deterioration are present, the sling shall be immediately removed from service.

(e)(4) Proof Testing

The employer shall ensure that before use, each new, repaired, or reconditioned alloy steel chain sling, including all welded components in the sling assembly, shall be proof tested by the sling manufacturer or equivalent entity, in accordance with paragraph 5.2 of the American Society of

Testing and Materials Specification A391-65, which is incorporated by reference as specified in Sec. 1910.6 (ANSI G61.1-1968). The employer shall retain a certificate of the proof test and shall make it available for examination.

This paragraph 1910.184(e)(4) requires that when an alloy steel chain sling is assembled with components that require welding in assembly, the completed sling must be proof tested by the sling manufacturer or equivalent entity, before the sling is used. The proof testing of slings is the responsibility of the sling manufacturer or equivalent entity as delineated by this standard as well as sections (g)(5) and (i)(8)(ii). The employer shall retain a "certificate of proof test" and shall make it available for examination by OSHA compliance officers. It is believed that any proof load testing of workplace slings other than the manufacturer or equivalent entity is an unacceptable loading and would necessitate that the sling be taken out of service unless written permission to test is obtained from the sling manufacturer.

When an alloy steel chain sling is made up of welded components that were individually proof tested, and no further welding is required to assemble the sling, the assembled chain sling does not have to be proof tested. The sling manufacturer or equal entity assembling the sling shall attach a tag identification with appropriate information, and furnish an appropriate certificate to the purchaser or his representative which indicates the rated capacity. Proof testing is not required when the sling is made up of components not requiring welding to assemble. The capacity of the sling shall be no greater than the rated capacity of the weakest component.

(e)(5) Sling use.

Alloy steel chain slings shall not be used with loads in excess of the rated capacities prescribed in Table N-184-1 (5.F.1). Slings not included in this table shall be used only in accordance with the manufacturer's recommendations.

Users of slings must not exceed the sling manufacturer's specifications and requirements pertaining to use and loadings.

(e)(6) Safe operating temperatures.

Alloy steel chain slings shall be permanently removed from service if they are heated above 1000 deg. F. When exposed to service temperatures in excess of 600 deg. F, maximum working load limits permitted in Table N-184-1 shall be reduced in accordance with the chain or sling manufacturer's recommendations.

(e)(7)

Repairing and reconditioning alloy steel chain slings.

(e)(7)(i)

Worn or damaged alloy steel chain slings or attachments shall not be used until repaired. When welding or heat testing is performed, slings shall not be used unless repaired, reconditioned and proof tested by the sling manufacturer or an equivalent entity.

Table 4.F.1. (N-184-1) Rated Capacity (Working Load Limit), for Alloy Steel Chain Slings

		Double sling vertical angle (1)			Triple and quadruple sling (3) Vertical angle (1)		
Chain Size, Inches	Single Branch Sling – 90 degree Loading	30 deg. (60 deg.)	45 deg. (45 deg.)	60 deg. (30 deg.)	30 deg. (60 deg.)	45 deg. (45 deg.)	60 deg. (30 deg.)
1/4	3,250	5,650	4,550	3,250	8,400	6,800	4,900
3/8	6,600	11,400	9,300	6,600	17,000	14,000	9,900
1/2	11,250	19,500	15,900	11,250	29,000	24,000	17,000
5/8	16,500	28,500	23,300	16,500	43,000	35,000	24,500
3/4	23,000	39,800	32,500	23,000	59,500	48,500	34,500
7/8	28,750	49,800	40,600	28,750	74,500	61,000	43,000
1	38,750	67,100	5,800	38,750	101,000	82,000	58,000
1 1/8	44,500	77,000	63,000	44,500	115,500	94,500	66,500
1 1/4	57,600	99,500	61,000	57,500	149,000	121,500	86,000
1 3/8	67,000	116,000	94,000	67,000	174,000	141,000	100,500
1 1/2	80,000	138,000	112,900	80,000	207,000	169,000	119,500
1 3/4	100,000	172,000	140,000	100,000	258,000	210,000	150,000

Footnote (1) Rating of multileg slings adjusted for angle of loading measured as the included angle between the inclined leg and the vertical as shown in Figure N-184-5.

Footnote (2) Rating of multileg slings adjusted for angle of loading between the inclined leg and the horizontal plane of the load, as shown in Figure N-184-5.

Footnote (3) Quadruple sling rating is same as triple sling because normal lifting practice may not distribute load uniformly to all 4 legs.

Table 4.F.2. (N-184-2) Minimum Allowable Chain Size at Any Point of Link

Chain size, inches	Minimum allowable chain size, inches
1/4	13/64
3/8	19/64
½	25/64
5/8	31/64
3/4	19/32
7/8	45/64
1	13/16
1 1/8	29/32
1 1/4	1
1 3/8	1 3/32
1 ½	1 3/16
1 3/4	1 13/32

(e)(7)(ii)

Mechanical coupling links or low carbon steel repair links shall not be used to repair broken lengths of chain.

(e)(8) Effects of wear.

If the chain size at any point of any link is less than that stated in Table N-184-2 (5.F.2), the sling shall be removed from service.

(e)(9) Deformed attachments.

(e)(9)(i)

Alloy steel chain slings with cracked or deformed master links, coupling links or other components shall be removed from service.

(e)(9)(ii)

Slings shall be removed from service if hooks are cracked, have been opened more than 15 percent of the normal throat opening measured at the narrowest point or twisted more than 10 degrees from the plane of the unbent hook.

(f) Wire Rope Slings

(f)(1)

Sling use. Wire rope slings shall not be used with loads in excess of the rated capacities shown in Tables N-184-3 through N-184-14. Slings not included in these tables shall be used only in accordance with the manufacturer's recommendations.

(f)(2) Minimum sling lengths.

(f)(2)(i)

Cable laid and 6 X 19 and 6 X 37 slings shall have a minimum clear length of wire rope 10 times the component rope diameter between splices, sleeves or end fittings.

(f)(2)(ii)

Braided slings shall have a minimum clear length of wire rope 40 times the component rope diameter between the loops or end fittings.

(f)(2)(iii)

Cable laid grommets, strand laid grommets and endless slings shall have a minimum circumferential length of 96 times their body diameter.

(f)(3) Safe operating temperatures.

Fiber core wire rope slings of all grades shall be permanently removed from service if they are exposed to temperatures in excess of 200 deg. F. When non-fiber core wire rope slings of any grade are used at temperatures above 400 deg. F or below minus 60 deg. F, recommendations of the sling manufacturer regarding use at that temperature shall be followed.

Table 4.F.3. (N-184-3) Rated Capacities For Single Leg Slings

6x19 and 6x37 Classification Improved Plow Steel Grade Rope With Fiber Core (FC)										
Rope		Rated capacities, tons (2,000 lb)								
Diameter (Inches)	(Constr.)	Vertical			Choker			Vertical Basket[1]		
		HT	MS	S	HT	MS	S	HT	MS	S
1/4	6x19	0.49	0.51	0.55	0.37	0.38	0.41	0.99	1.0	1.1
5/16	6x19	0.76	0.79	0.85	0.57	0.59	0.64	1.5	1.6	1.7
3/8	6x19	1.1	1.1	1.2	0.80	0.85	0.91	2.1	2.2	2.4
7/16	6x19	1.4	1.5	1.6	1.1	1.1	1.2	2.9	3.0	3.3
½	6x19	1.8	2.0	2.1	1.4	1.5	12.6	3.7	3.9	4.3
9/16	6x19	2.3	2.5	2.7	1.7	1.9	2.0	4.6	5.0	5.4
5/8	6x19	2.8	3.1	3.3	2.1	2.3	2.5	5.6	6.2	6.7
3/4	6x19	3.9	4.4	4.8	2.9	3.3	3.6	7.8	8.8	9.5
7/8	6x19	5.1	5.9	6.4	3.9	4.5	4.8	10.0	12.0	13.0
1	6x19	6.7	7.7	8.4	5.0	5.8	6.3	13.0	15.0	17.0
1 1/8	6x19	8.4	9.5	10.0	6.3	7.1	7.9	17.0	19.0	21.0
1 1/4	6x37	9.8	11.0	12.0	7.4	8.3	9.2	21.0	22.0	25.0
1 3/4	6x37	12.0	13.0	15.0	8.9	10.0	11.0	24.0	27.0	30.0
1 ½	6x37	14.0	16.0	15.0	10.0	12.0	13.0	28.0	32.0	35.0
1 5/8	6x37	16.0	18.0	21.0	12.0	14.0	15.0	33.0	27.0	41.0
1 3/4	6x37	19.0	21.0	24.0	14.0	16.0	18.0	38.0	43.0	48.0
2	6x37	25.0	28.0	31.0	18.0	21.0	23.0	49.0	55.0	62.0

HT = Hand Tucked Splice and Hidden Tuck Splice. For hidden tuck splice (IWRC) use values in HT columns.

MS = Mechanical Splice.

S = Swaged or Zinc Poured Socket.

Footnote (1) These values only apply when the D/d ratio for HT slings is 10 or greater, and for MS and S slings is 20 or greater where: D = Diameter of curvature around which the body of the sling is bent; d = Diameter of rope.

Table 4.F.4. (N-184-4) Rated Capacities For Single Leg Slings

6x19 and 6x37 Classification Improved Plow Steel Grade Rope With Independent Wire Rope Core (IWRC)										
Rope		Rated capacities, tons (2,000 lb)								
Diameter (Inches)	(Constr.)	Vertical			Choker			Vertical Basket[1]		
		HT	MS	S	HT	MS	S	HT	MS	S
1/4	6x19	0.53	0.56	0.59	0.40	0.42	0.44	1.0	1.1	1.2
5/16	6x19	0.81	0.87	0.92	0.61	0.65	0.69	1.6	1.7	1.8
3/8	6x19	1.1	1.2	1.3	0.86	0.93	0.98	2.3	2.5	2.6
7/16	6x19	1.5	1.7	1.8	1.2	1.3	1.3	3.1	3.4	3.5
½	6x19	2.0	2.2	2.3	1.5	1.6	1.7	3.9	4.4	4.6
9/16	6x19	2.5	2.7	2.9	1.8	2.1	2.2	4.9	5.5	5.8
5/8	6x19	3.0	3.4	3.6	2.2	2.5	2.7	6.0	6.8	7.2
3/4	6x19	4.2	4.9	5.1	3.1	3.6	3.8	8.4	9.7	10.0
7/8	6x19	5.5	6.6	6.9	4.1	4.9	5.2	11.0	13.0	14.0
1	6x19	7.2	8.5	9.0	5.4	6.4	6.7	14.0	17.0	18.0
1 1/8	6x19	9.0	10.0	11.0	6.8	7.8	8.5	28.0	21.0	23.0
1 1/4	6x37	10.0	12.0	13.0	7.9	9.2	9.9	21.0	24.0	26.0
1 3/4	6x37	13.0	15.0	16.0	9.6	11.0	12.0	25.0	29.0	32.0
1 ½	6x37	15.0	17.0	19.0	11.0	13.0	14.0	30.0	35.0	38.0
1 5/8	6x37	18.0	20.0	22.0	13.0	15.0	17.0	35.0	41.0	44.0
1 3/4	6x37	20.0	24.0	26.0	15.0	18.0	19.0	41.0	47.0	51.0
2	6x37	26.0	30.0	33.0	20.0	23.0	25.0	53.0	61.0	66.0

HT = Hand Tucked Splice. For hidden tuck splice (IWRC) use Table 1 values in HT column.

MS = Mechanical Splice.

S = Swaged or Zinc Poured Socket.

Footnote (1) These values only apply when the D/d ratio for HT slings is 10 or greater, and for MS and S slings is 20 or greater where: D = Diameter of curvature around which the body of the sling is bent; d = Diameter of rope.

Table 4.F.5. (N-184-5) Rated Capacities For Single Leg Slings

Cable Laid Rope -- Mechanical Splice Only 7x7x7 & 7x19 Constructions Galvanized Aircraft Grade Rope 7x6x19 IWRC Construction Improved Plow Steel Grade Rope				
Rope		Rated Capacities, tons (2000 lb)		
Dia (inches)	Constr	Vertical	Choker	Vertical basket[1]
1/4	7x7 x7	0.50	0.38	1.0
3/8	7x7 x7	1.1	0.81	2.0
½	7x7 x7	1.8	1.4	3.7
5/8	7x7 x7	2.8	2.1	5.5
3/4	7x7 x7	3.8	2.9	7.6
5/8	7x7 x19	2.9	2.2	5.8
3/4	7x7 x19	4.1	3.0	8.1
7/8	7x7 x19	5.4	4.0	11.0
1	7x7 x19	6.9	5.1	14.0
1 1/8	7x7 x19	8.2	6.2	16.0
1 1/4	7x7 x19	9.9	7.4	20.0
3/4	7x6x19 IWRC	3.8	2.8	7.6
7/8	7x6x19 IWRC	5.0	3.8	10.0
1	7x6x19 IWRC	6.4	4.8	13.0
1 1/8	7x6x19 IWRC	7.7	5.9	15.0
1 1/4	7x6x19 IWRC	9.2	6.9	18.0
1 5/16	7x6x19 IWRC	10.0	7.5	20.0
1 3/8	7x6x19 IWRC	11.0	8.2	22.0
1 1/2	7x6x19 IWRC	13.0	9.6	26.0

Footnote (1) These values only apply when the D/d ratio is 10 or greater where: D = Diameter of curvature around which the body of the sling is bent; d = Diameter of rope.

Table 4.F.6. (N-184-6) Rated Capacities For Single Leg Slings

8-Part and 6-Part Braided Rope 6x7 and 6x19 Construction Improved Plow Steel Grade Rope 7x7 Construction Galvanized Aircraft Grade Rope							
Component ropes		Rated Capacities, tons (2,000 lb)					
Diameter (inches)	Constr	Vertical		Choker		Basket, vertical to 30⁰ ¹	
		8-Part	6-Part	8-Part	6-Part	8-Part	6-Part
3/32	6x7	0.42	0.32	0.32	0.24	0.74	0.55
1/8	6x7	0.75	0.57	0.57	0.42	1.3	0.98
3/16	6x7	1.7	1.3	1.3	0.94	2.9	2.2
3/32	7x7	0.51	0.39	0.38	0.29	0.89	0.67
1/8	7x7	0.95	0.7	0.71	0.71	0.53	1.2
3/16	7x7	2.1	1.5	1.5	1.5	1.2	2.7
3/16	6x19	1.7	1.3	1.3	1.3	0.98	2.2
1/4	6x19	3.1	2.3	2.3	2.3	1.7	4.0
5/16	6x19	4.8	3.6	3.6	3.6	2.7	6.2
3/8	6x19	6.8	5.1	5.1	5.1	3.8	8.9
7/16	6x19	9.3	6.9	6.0	6.9	5.2	12.0
½	6x19	12.0	9.0	9.0	9.0	6.7	15.0
9/16	6x19	15.0	11.0	11.0	11.0	8.5	20.0
5/8	6x19	19.0	14.0	14.0	14.0	10.0	24.0
3/4	6x19	27.0	20.0	20.0	20.0	15.0	35.0
7/8	6x19	36.0	27.0	27.0	27.0	20.0	47.0
1	6x19	47.0	35.0	35.0	35.0	26.0	61.0

Footnote (1) These values only apply when the D/d ratio is 20 or greater where: D = Diameter of curvature around which the body of the sling is bent; d = Diameter of component rope.

Table 4.F.7. (N-184-7) Rated Capacities For 2-Leg and 3-Leg Bridle Slings

Rope		6x19 and 6x37 Classification Improved Plow Steel Grade Rope With Fiber Core (FC) [Horizontal angles shown in parentheses]											
		Rated Capacities, tons (2,000 lb)											
		2-Leg bridle slings						3-Leg bridle slings					
Dia (in.)	Constr	30° (60°)		45° angle		60° (30°)		30° (60°)		45° angle		60° (30°)	
		HT	MS	HT	MS	HT	MS	HT	MS	HT	MS	HT	MS
1/4	6x19	0.85	0.83	0.70	0.72	0.49	0.51	1.3	1.3	1.0	1.1	0.74	0.76
5/16	6x19	1.3	1.4	1.1	1.1	0.76	0.79	2.0	2.0	1.6	1.7	1.1	1.2
3/8	6x19	1.8	1.9	1.5	1.6	1.1	1.1	2.8	2.9	2.3	2.4	1.6	1.7
7/16	6x19	2.5	2.6	2.0	2.2	1.4	1.5	3.7	4.0	3.0	3.2	2.1	2.3
½	6x19	3.2	3.4	2.6	2.8	1.8	2.0	4.8	5.1	3.9	4.2	2.8	3.0
9/16	6x19	4.0	4.3	3.2	3.5	2.3	2.5	6.0	6.5	4.9	5.3	3.4	3.7
5/8	6x19	4.8	5.3	4.0	4.4	2.8	3.1	7.3	8.0	5.9	6.5	4.2	4.6
3/4	6x19	6.8	7.6	5.5	6.2	3.9	4.4	10.0	11.0	8.3	9.3	5.8	6.6
7/8	6x19	8.9	10.0	7.3	8.4	5.1	5.9	13.0	15.0	11.0	13.0	7.7	8.9
1	6x19	11.0	13.0	9.4	11.0	6.7	7.7	17.0	20.0	14.0	16.0	10.0	11.0
1 1/8	6x19	14.0	16.0	12.0	13.0	8.4	9.3	22.0	24.0	18.0	20.0	13.0	14.0
1 1/4	6x37	17.0	19.0	14.0	16.0	9.8	11.0	25.0	29.0	21.0	23.0	15.0	17.0
1 3/8	6x37	20.0	23.0	17.0	19.0	12.0	13.0	31.0	35.0	25.0	28.0	18.0	20.0
1 ½	6x37	24.0	27.0	20.0	22.0	14.0	16.0	36.0	41.0	30.0	33.0	21.0	24.0
1 5/8	6x37	28.0	32.0	23.0	26.0	16.0	18.0	43.0	48.0	35.0	39.0	25.0	28.0
1 3/4	6x37	33.0	37.0	27.0	30.0	19.0	21.0	49.0	56.0	40.0	45.0	28.0	32.0
2	6x37	43.0	48.0	35.0	39.0	25.0	28.0	64.0	72.0	52.0	59.0	37.0	41.0

HT = Hand Tucked Splice.

MS = Mechanical Splice.

Table 4.F.8. (N-184-8) Rated Capacities For 2-Leg and 3-Leg Bridle Slings

6x19 and 6x37 Classification Improved Plow Steel Grade Rope With Fiber Core (FC) [Horizontal angles shown in parentheses]														
Rope		Rated Capacities, tons (2,000 lb)												
		2-Leg bridle slings						3-Leg bridle slings						
Dia (in.)	Constr	30^0 (60^0)		45^0 angle		60^0 (30^0)		30^0 (60^0)		45^0 angle		60^0 (30^0)	
		HT	MS	HT	MS	HT	MS	HT	MS	HT	MS	HT	MS
1/4	6x19	0.85	0.83	0.70	0.72	0.49	0.51	1.3	1.3	1.0	1.1	0.74	0.76
5/16	6x19	1.3	1.4	1.1	1.1	0.76	0.79	2.0	2.0	1.6	1.7	1.1	1.2
3/8	6x19	1.8	1.9	1.5	1.6	1.1	1.1	2.8	2.9	2.3	2.4	1.6	1.7
7/16	6x19	2.5	2.6	2.0	2.2	1.4	1.5	3.7	4.0	3.0	3.2	2.1	2.3
½	6x19	3.2	3.4	2.6	2.8	1.8	2.0	4.8	5.1	3.9	4.2	2.8	3.0
9/16	6x19	4.0	4.3	3.2	3.5	2.3	2.5	6.0	6.5	4.9	5.3	3.4	3.7
5/8	6x19	4.8	5.3	4.0	4.4	2.8	3.1	7.3	8.0	5.9	6.5	4.2	4.6
3/4	6x19	6.8	7.6	5.5	6.2	3.9	4.4	10.0	11.0	8.3	9.3	5.8	6.6
7/8	6x19	8.9	10.0	7.3	8.4	5.1	5.9	13.0	15.0	11.0	13.0	7.7	8.9
1	6x19	11.0	13.0	9.4	11.0	6.7	7.7	17.0	20.0	14.0	16.0	10.0	11.0
1 1/8	6x19	14.0	16.0	12.0	13.0	8.4	9.3	22.0	24.0	18.0	20.0	13.0	14.0
1 1/4	6x37	17.0	19.0	14.0	16.0	9.8	11.0	25.0	29.0	21.0	23.0	15.0	17.0
1 3/8	6x37	20.0	23.0	17.0	19.0	12.0	13.0	31.0	35.0	25.0	28.0	18.0	20.0
1 ½	6x37	24.0	27.0	20.0	22.0	14.0	16.0	36.0	41.0	30.0	33.0	21.0	24.0
1 5/8	6x37	28.0	32.0	23.0	26.0	16.0	18.0	43.0	48.0	35.0	39.0	25.0	28.0
1 3/4	6x37	33.0	37.0	27.0	30.0	19.0	21.0	49.0	56.0	40.0	45.0	28.0	32.0
2	6x37	43.0	48.0	35.0	39.0	25.0	28.0	64.0	72.0	52.0	59.0	37.0	41.0

HT = Hand Tucked Splice.

MS = Mechanical Splice.

Table 4.F.9. (N-184-9) Rated Capacities For 2-Leg and 3-Leg Bridle Slings

Cable Laid Rope -- Mechanical Splice Only
7x7x7 & 7x19 Constructions Galvanized Aircraft Grade Rope
7x6x19 IWRC Construction Improved Plow Steel Grade Rope
(Horizontal angles shown in parentheses)

| Rope | | Rated Capacities, tons (2000 lb) | | | | | |
| Dia (inches) | Constr | 2-Leg bridle sling | | | 3-Leg bridle sling | | |
		30° (60°)	45° angle	60° (30°)	30° (60°)	45° angle	60° (30°)
1/4	7x7x7	0.87	0.71	0.50	1.3	1.1	0.75
3/8	7x7x7	1.9	1.5	1.1	2.8	2.3	1.6
1/2	7x7x7	3.2	2.6	1.8	4.8	3.9	2.8
5/8	7x7x7	4.8	3.9	2.8	7.2	5.9	4.2
3/4	7x7x7	6.6	5.4	3.8	9.9	8.1	3.7
5/8	7x7x19	5.0	4.1	2.9	7.5	6.1	4.3
3/4	7x7x19	7.0	5.7	4.1	10.0	8.6	6.1
7/8	7x7x19	9.3	7.6	5.4	14.0	11.0	8.1
1	7x7x19	12.0	9.7	6.9	18.0	14.0	10.0
1 1/8	7x7x19	14.0	12.0	8.2	21.0	17.0	12.0
1 1/4	7x7x19	17.0	14.0	9.9	26.0	21.0	15.0
3/4	7x6x19 IWRC	6.6	5.4	3.8	9.9	8.0	5.7
7/8	7x6x19 IWRC	8.7	7.1	5.0	13.0	11.0	7.5
1	7x6x19 IWRC	11.0	9.0	6.4	17.0	13.0	9.6
1 1/8	7x6x19 IWRC	13.0	11.0	7.7	20.0	16.0	11.0
1 1/4	7x6x19 IWRC	16.0	13.0	9.2	24.0	20.0	14.0
1 5/16	7x6x19 IWRC	17.0	14.0	10.0	26.0	21.0	15.0
1 3/8	7x6x19 IWRC	19.0	15.0	11.0	28.0	23.0	16.0
1 1/2	7x6x19 IWRC	22.0	18.0	13.0	33.0	27.0	19.0

Table 4.F.10. (N-184-10) Rated Capacities For 2-Leg and 3-Leg Bridle Slings

8-Part and 6-Part Braided Rope
6 x 7 and 6 x 19 Construction Improved Plow Steel Grade Rope
7 x 7 Construction Galvanized Aircraft Grade Rope
[Horizontal angles shown in parentheses]

Rated Capacities, tons (2,000 lb)

Rope		2-Leg bridle slings						3-Leg bridle sling					
		30°(60°)		45° angle		60°(30°)		30°(60°)		45° angle		60°(30°)	
Dia (in.)	Constr	8-Part	6-Part	8-Part	6-Part	8-Part	6-Part	8-Part	6-Part	8-Part	6-Part	8-Part	6-Part
3/32	6x7	0.74	0.55	0.60	0.45	0.42	0.32	1.1	0.83	0.90	0.68	0.64	0.48
1/8	6x7	1.3	0.98	1.1	0.80	0.76	0.57	2.0	1.5	1.6	1.2	1.1	0.85
3/16	6x7	2.9	2.2	2.4	1.8	1.7	1.3	4.4	3.3	3.6	2.7	2.5	1.9
3/32	7x7	0.89	0.67	0.72	0.55	0.51	0.39	1.3	1.0	1.1	0.82	0.77	0.58
1/8	7x7	1.6	1.2	1.3	1.0	0.95	0.71	2.5	1.8	2.0	1.5	1.4	1.1
3/16	7x7	3.6	2.7	2.9	2.2	2.1	1.5	5.4	4.0	4.4	3.3	3.1	2.3
3/16	6x19	3.0	2.2	2.4	1.8	1.7	1.3	4.5	3.4	3.7	2.8	2.6	1.9
1/4	6x19	5.3	4.0	4.3	3.2	3.1	2.3	8.0	6.0	6.5	4.9	4.6	3.4
5/16	6x19	8.3	6.2	6.7	5.0	4.8	3.6	12.0	9.3	10.0	7.6	7.1	5.4
3/8	6x19	12.0	8.9	9.7	7.2	6.8	5.1	18.0	13.0	14.0	11.0	10.0	7.7
7/16	6x19	16.0	12.0	13.0	9.8	9.3	6.9	24.0	18.0	20.0	15.0	14.0	10.0
1/2	6x19	21.0	155.0	17.0	13.0	12.0	9.0	31.0	23.0	25.0	19.0	18.0	13.0
9/16	6x19	25.0	20.0	21.0	16.0	15.0	11.0	39.0	29.0	32.0	243.0	23.0	17.0
5/8	6x19	32.0	24.0	26.0	20.0	10.0	14.0	48.0	36.0	40.0	30.0	28.0	21.0
3/4	6x19	46.0	35.0	38.0	28.0	27.0	20.0	69.0	52.0	56.0	42.0	40.0	30.0
7/8	6x19	62.0	47.0	51.0	38.0	36.0	27.0	94.0	70.0	76.0	57.0	54.0	40.0
1	6x19	81.0	61.0	66.0	50.0	47.0	35.0	122.0	91.0	99.0	74.0	70.0	53.0

Table 4.F.11. (N-184-11) Rated Capacities For Strand Laid Grommet – Hand Tucked

Improved Plow Steel Grade Rope				
Rope body		Rated capacities, tons (2,000lb)		
Dia (inches	Constr	Vertical	Choker	Vertical basket[1]
1/4	7x19	0.85	0.64	1.7
5/16	7x19	1.3	1.0	2.6
3/8	7x19	1.9	1.4	3.8
7/16	7x19	2.6	1.9	5.2
½	7x19	3.3	2.5	6.7
9/16	7x19	4.2	3.1	8.4
5/8	7x19	5.2	3.9	10.0
3/4	7x19	7.4	5.6	15.0
7/8	7x19	10.0	7.5	20.0
1	7x19	13.0	9.7	26.0
1 1/8	7x19	16.0	12.0	32.0
1 1/4	7x37	18.0	14.0	37.0
1 3/8	7x37	22.0	16.0	44.0
1 1/2	7x37	26.0	19.0	52.0

Footnote (1) These values only apply when the D/d ratio is 5 or greater where:
D = Diameter of curvature around which rope is bent. d =Diameter of rope body.

Table 4.F.12. (N-184-12) Rated Capacities For Cable Laid Grommet – Hand Tucked

7 x 6 x 7 and 7 x 6 x 19 Construction Improved Plow Steel Rope 7 x 7 x 7 Construction Galvanized Aircraft Grade Rope				
Cable body		Rated capacities, tons (2,000lb)		
Dia (inches	Constr	Vertical	Choker	Vertical basket[1]
3/8	7x6x7	1.3	0.95	2.5
9/16	7x6x7	2.8	2.1	5.6
5/8	7x6x7	3.8	2.8	7.6
3/8	7x7x7	1.6	1.2	3.2
9/16	7x7x7	3.5	2.6	6.9
5/8	7x7x7	4.5	3.4	9.0
5/8	7x6x19	3.9	3.0	7.9
3/4	7x6x19	5.1	3.8	10.0
1 5/16	7x6x19	7.9	5.9	16.0
1 1/8	7x6x19	11.0	8.4	22.0
1 5/16	7x6x19	15.0	11.0	30.0
1 ½	7x6x19	19.0	14.0	39.0
1 11/16	7x6x19	24.0	18.0	49.0
1 7/8	7x6x19	30.0	22.0	60.0
2 1/4	7x6x19	42.0	31.0	84.0
2 5/8	7x6x19	56.0	42.0	112.0

Footnote (1) These values only apply when the D/d ratio is 5 or greater where:
D = Diameter of curvature around which cable body is bent. d =Diameter of cable body.

Table 4.F.13. (N-184-13) Rated Capacities For Strand Endless Slings – Mechanical Joints

Improved Plow Steel Grade Rope				
Cable body		Rated capacities, tons (2,000lb)		
Dia (inches	Constr	Vertical	Choker	Vertical basket[1]
1/4	6x19[2]	0.92	0.69	1.8
3/8	6x19[2]	2.0	1.5	4.1
½	6x19[2]	3.6	2.7	7.2
5/8	6x19[2]	5.6	4.2	11.0
3/4	6x19[2]	8.0	6.0	16.0
7/8	6x19[2]	11.0	8.1	21.0
1	6x19[2]	14.0	10.0	28.0
1 1/8	6x19[2]	18.0	13.0	35.0
1 1/4	6x37[2]	21.0	15.0	41.0
1 3/8	6x37[2]	25.0	19.0	50.0
1 1/2	6x37[2]	29.0	22.0	59.0

Footnote (1) These values only apply when the D/d ratio is 5 or greater where:
D =Diameter of curvature around which rope is bent. d =Diameter of rope body.
Footnote (2) IWRC.

Table 4.F.14. (N-184-14) Rated Capacities For Cable Laid Endless Slings – Mechanical Joints

7x7 x 7 and 7 x 7 x 19 Construction Galvanized Aircraft Grade Rope 7 X 6 X 19 IWRC Construction Improved Plow Steel Grade Rope				
Cable body		Rated capacities, tons (2,000lb)		
Dia (inches)	Constr	Vertical	Choker	Vertical basket[1]
1/4	7x7x7	0.83	0.62	1.6
3/8	7x7x7	1.8	1.3	3.5
½	7x7x7	3.0	2.3	6.1
5/8	7x7x7	4.5	3.4	9.1
3/4	7x7x7	6.3	4.7	12.0
5/8	7x7x19	4.7	3.5	9.5
3/4	7x7x19	6.7	5.0	13.0
7/8	7x7x19	8.9	6.6	18.0
1	7x7x19	11.0	8.6\5	22.0
1 1/8	7x7x19	14.0	10.0	28.0
1 1/4	7x7x19	17.0	12.0	33.0
3/4	7x6x19[2]	6.2	4.7	12.0
7/8	7x6x19[2]	8.3	6.2	16.0
1	7x6x19[2]	10.0	7.9	21.0
1 1/8	7x6x19[2]	13.0	9.7	26.0
1 1/4	7x6x19[2]	16.0	12.0	31.0
1 3/8	7x6x19[2]	18.0	14.0	37.0
1 1/2	7x6x19[2]	22.0	16.0	43.0

Footnote (1) These values only apply when the D/d value is 5 or greater where:
D = Diameter of curvature around which cable body is bent. d =Diameter of cable body.
Footnote (2) IWRC.

(f)(4) End attachments.

(f)(4)(i)

Welding of end attachments, except covers to thimbles, shall be performed prior to the assembly of the sling.

(f)(4)(ii)

All welded end attachments shall not be used unless proof tested by the manufacturer or equivalent entity at twice their rated capacity prior to initial use. The employer shall retain a certificate of the proof test, and make it available for examination.

> *General purpose wire ropes are used for a broad range of applications and are available in different standard classifications (based on the number of strands and wires per strand): i.e., 6x19, and 6x37. Within each standard classification, ropes of the same size, grade, and core have the same strengths and weights. The 6x19 classification ropes provide an excellent balance between fatigue and wear resistance. They give excellent service with sheaves and drums of moderate size. The 6x19 classification ropes contain 6 strands with 15 through 26 wires per strand, no more than 12 of which are outside wires. The 6x37 classification wire ropes are more flexible but less abrasion resistant than the 6x19 classification. Each strand contains numerous small diameter wires. As the number of wires increases, flexibility increases. The 6x37 classification ropes contain 6 strands with 27 through 49 wires, no more than 18 of which are outside wires.*

(f)(5) Removal from service.

Wire rope slings shall be immediately removed from service if any of the following conditions are present:

(f)(5)(i)

Ten randomly distributed broken wires in one rope lay, or five broken wires in one strand in one rope lay.

(f)(5)(ii)

Wear or scraping of one-third the original diameter of outside individual wires.

> *The OSHA standards 1910.184(f)(5)(i) and 29 CFR 1910.184(f)(5)(ii) require wire rope slings to be removed from service immediately when the following conditions are found:*
>
> *(i) Ten randomly distributed broken wires in one rope lay, or five broken wires in one strand in one rope lay.*
>
> *(ii) Wear and scraping of one-third the original diameter of outside individual wires. Compliance with 1910.184(f)(5)(i) is determined by inspection of the rope sling.*

The following method may be used to determine whether the wire rope sling must be removed from service as required by 1910.184(f)(5)(ii). The outside individual wires are not separated from the wire rope to make them available for measuring. Use a micrometer to measure the wear or scraping of the wire rope. Compare the original diameter of whole wire rope to that of the worn area. If the difference of this measurement is equal to, or greater than, one-third the original diameter of an individual outside wire, the wire rope sling must be removed from service.

OSHA will allow a wire rope to be left in service with respect to a pass/fail gage measurement if the difference between the original diameter of the whole wire rope and a pass/fail gage OD failed measurement is less than one-third the original diameter of the outside individual wire.

Slings and all fastenings and attachments must be inspected for damage or defects each day before being used by a competent person designated by the employer. Where service conditions warrant, additional inspections must be performed during sling use. Damaged or defective slings must be immediately removed from service.

(f)(5)(iii)

Kinking, crushing, bird caging or any other damage resulting in distortion of the wire rope structure.

(f)(5)(iv)

Evidence of heat damage.

(f)(5)(v)

End attachments that are cracked, deformed or worn.

(f)(5)(vi)

Hooks that have been opened more than 15 percent of the normal throat opening measured at the narrowest point or twisted more than 10 degrees from the plane of the unbent hook.

(f)(5)(vii)

Corrosion of the rope or end attachments.

(g) Metal Mesh Slings

(g)(1)

Sling marking. Each metal mesh sling shall have permanently affixed to it a durable marking that states the rated capacity for vertical basket hitch and choker hitch loadings.

(g)(2) Handles.

Handles shall have a rated capacity at least equal to the metal fabric and exhibit no deformation after proof testing.

(g)(3) Attachments of handles to fabric.

The fabric and handles shall be joined so that:

(g)(3)(i)

The rated capacity of the sling is not reduced.

(g)(3)(ii)

The load is evenly distributed across the width of the fabric.

(g)(3)(iii)

Sharp edges will not damage the fabric.

(g)(4)

Sling coatings. Coatings which diminish the rated capacity of a sling shall not be applied.

(g)(5)

Sling testing. All new and repaired metal mesh slings, including handles, shall not be used unless proof tested by the manufacturer or equivalent entity at a minimum of 1 ½ times their rated capacity. Elastomer impregnated slings shall be proof tested before coating.

> The requirements imposed by any company upon their subcontractors to repeatedly proof test slings to more than 1 1/2 times a rated working load is not a recognized inspection procedure under the OSHA standards and would be a violation of 1910.184(c)(4). However, should the sling manufacturer provide written permission to test the slings on a regular basis to a load greater than the designated working load, OSHA could consider the violation de minimis. Of course the procedures for such testing would also need to comply with the manufacturer's recommendations.

> The repeated testing of equipment seems to provide little more assurance of continued reliability than the visual inspections required by the various standards, including OSHA's. Properly conducted visual inspections together with the careful reporting of misuse and various damaging exposures will provide for the continued reliability of the lifting equipment. Should periodic load testing be desired, it is recommended that slings be returned to the manufacturer or equivalent entity for the conduct of detailed inspection and load tests. Only the manufacturer or equivalent entity is permitted to proof test and certify. Certification is exclusively the right of only the manufacturer of new slings. Repaired slings may be certified by an equivalent entity that made the repairs.

> An "equivalent entity" is defined by the general industry standard 29 CFR 1910.184(b), which states: "A person or organization (including an employer) which, by possession of equipment, technical knowledge and skills, can perform with equal competence the same repairs and tests as the person or organization with which it is equated."

In general, manufacturers of chains and their designated service representatives are considered to be "equivalent entities" with regard to making repairs of alloy steel chains. Manufacturers of alloy steel chains often designate "equivalent entities" under a licensing arrangement between the parties. The various methods by which alloy steel chains are repaired and/or refurbished are very complex and require extensive equipment as well as highly skilled personnel. Any company may approach the manufacturer to establish an agreement, and where such an agreement is in force and the work accomplished in accord with the manufacturer's instructions, OSHA would consider this to be in compliance with the standard.

Slings made with chains are required to be in compliance with the manufacturer's specifications, the standards, or may be specifically constructed to specifications prepared by a licensed professional engineer familiar with the technology and authorized to develop such specifications.

(g)(6)

Proper use of metal mesh slings. Metal mesh slings shall not be used to lift loads in excess of their rated capacities as prescribed in Table N-184-15. Slings not included in this table shall be used only in accordance with the manufacturer's recommendations.

(g)(7) Safe operating temperatures.

Metal mesh slings which are not impregnated with elastomers may be used in a temperature range from minus 20° F to plus 550° F without decreasing the working load limit. Metal mesh slings impregnated with polyvinyl chloride or neoprene may be used only in a temperature range from zero degrees to plus 200° F. For operations outside these temperature ranges or for metal mesh slings impregnated with other materials, the sling manufacturer's recommendations shall be followed.

(g)(8) Repairs

(g)(8)(i)

Metal mesh slings which are repaired shall not be used unless repaired by a metal mesh sling manufacturer or an equivalent entity.

(g)(8)(ii)

Once repaired, each sling shall be permanently marked or tagged, or a written record maintained, to indicate the date and nature of the repairs and the person or organization that performed the repairs. Records of repairs shall be made available for examination.

(g)(9)

Removal from service. Metal mesh slings shall be immediately removed from service if any of the following conditions are present:

(g)(9)(i)

A broken weld or broken brazed joint along the sling edge.

Table 4.F.15. (N-184-15) Rated Capacities

Sling width in inches	Vertical or choker	Vertical basket	Effect of angle on rated capacities in basket hitch		
			30^0 (60^0)	45^0 (45^0)	60^0 (30^0)
Heavy Duty - 10 Ga 35 Spirals/Ft of sling width					
2	1,500	3,000	2,600	2,100	1,500
3	2,700	5,400	4,700	3,800	2,700
4	4,000	8,000	6,900	5,600	4,000
6	6,000	12,000	10,400	8,400	6,000
8	8,000	16,000	13,800	11,300	8,000
10	10,000	20,000	17,000	14,100	10,000
12	12,000	24,000	20,700	16,900	12,000
14	14,000	28,000	24,200	19,700	14,000
16	16,000	32,000	27,700	22,600	16,000
18	18,000	36,000	31,100	25,400	18,000
20	20,000	40,000	34,600	28,200	20,000
Medium Duty - 12 Ga 43 Spirals/Ft of sling width					
2	1,350	2,700	2,300	1,900	1,400
3	2,000	4,000	3,500	2,800	2,000
4	2,700	5,400	4,700	3,800	2,700
6	4,500	9,000	7,800	6,400	4,500
8	6,000	12,000	10,400	8,500	6,000
10	7,500	15,000	13,000	10,600	7,500
12	9,000	18,000	15,600	12,700	9,000
14	10,500	21,000	18,200	14,800	10,500
16	12,000	24,000	20.800	17,000	12,000
18	13,500	27,000	23,400	19,100	13,500
20	15,000	30,000	26,000	21,200	15,000
Light Duty - 14 Ga 59 Spirals/Ft of sling width					
2	900	1,800	1,600	1,300	900
3	1,400	2,800	2,400	2,000	1,400
4	2,000	4,000	3,500	2,800	2,000
6	3,000	6,000	5,200	4,200	3,000
8	4,000	8,000	6,900	5,700	4,000
10	5,000	10,000	8,600	7,100	5,000
12	6,000	12,000	10,400	8,500	6,000
14	7,000	14,000	12,100	9,900	7,000
16	8,000	16,000	13,900	11,300	8,000
18	9,000	18,000	15,600	12,700	9,000
20	10,000	20,000	17,300	14,100	10,000

Carbon Steel and Stainless Steel Metal Mesh slings
[Horizontal angles shown in parentheses]

(g)(9)(ii)

Reduction in wire diameter of 25 per cent due to abrasion or 15 per cent due to corrosion.

(g)(9)(iii)

Lack of flexibility due to distortion of the fabric.

(g)(9)(iv)

Distortion of the female handle so that the depth of the slot is increased more than 10 per cent.

(g)(9)(v)

Distortion of either handle so that the width of the eye is decreased more than 10 per cent.

(g)(9)(vi)

A 15 percent reduction of the original cross sectional area of metal at any point around the handle eye.

(g)(9)(vii)

Distortion of either handle out of its plane.

(h) Natural and Synthetic Fiber Rope Slings

(h)(1)

Sling use.

(h)(1)(i)

Fiber rope slings made from conventional three strand construction fiber rope shall not be used with loads in excess of the rated capacities prescribed in Tables N-184-16 through N-184-19.

(h)(1)(ii)

Fiber rope slings shall have a diameter of curvature meeting at least the minimums specified in Figs. N-184-4 and N-184-5.

(h)(1)(iii)

Slings not included in these tables shall be used only in accordance with the manufacturer's recommendations.

(h)(2)

Safe operating temperatures. Natural and synthetic fiber rope slings, except for wet frozen slings, may be used in a temperature range from minus 20 deg. F to plus 180 deg. F without decreasing the working load limit. For operations outside this temperature range and for wet frozen slings, the sling manufacturer's recommendations shall be followed.

Table 4.F.16. (N-184-16) Manila Rope Slings

[Angle of rope to vertical shown in parentheses]

Rope dia. nominal in inches	Nominal wt. per 100 ft in pounds	Eye and eye sling						Endless sling					
		Verti-cal hitch	Choker hitch	Basket hitch; Angle of rope to horizontal				Verti-cal hitch	Choker hitch	Basket hitch; Angle of rope to horizontal			
				90° (0°)	60 (30°)	45 (45°)	30 (60°)			90° (0°)	60 (30°)	45 (45°)	30 (60°)
½	7.5	480	240	960	830	680	480	865	430	1,730	1,500	1,220	865
9/16	10.4	620	310	1,240	1,070	875	620	1,120	560	2,230	1,930	1,580	1,120
5/8	13.3	790	395	1,580	1,370	1,120	790	1,420	710	2,840	2,460	2,010	1,420
3/4	16.7	970	485	1,940	1,680	1,370	970	1,750	875	3,490	3,020	2,470	1,750
13/16	19.5	1,170	585	2,340	2,030	1,650	1,170	2,110	1,050	4,210	3,650	2,980	2,110
7/8	22.5	1,390	695	2,780	2,410	1,970	1,390	2,500	1,250	5,000	4,330	3,540	2,500
1	27.0	1,620	810	3,240	2,810	2,290	1,620	2,920	1,460	5,830	5,050	4,120	2,920
1 1/16	31.3	1,890	945	3,780	3,270	2,670	1,890	3,400	1,700	6,800	5,890	4,810	3,400
1 1/8	36.0	2,160	1,080	4,320	3,740	3,050	2,160	3,890	1,940	7,780	6,730	5,500	3,890
1 1/4	41.7	2,430	1,220	4,860	4,210	3,440	2,430	4,370	2,190	8,750	7,580	6,190	4,370
1 5/16	47.9	2,700	1,350	5,400	4,680	3,820	2,700	4,860	2,430	9,720	8,420	6,870	4,860
1 ½	59.9	3,330	1,670	6,660	5,770	4,710	3,330	5,990	3,000	12,000	10,400	8,480	5,990
1 5/8	74.6	4,050	2,030	8,100	7,010	5,730	4,050	7,290	3,650	14,600	12,600	10,300	7,290
1 3/4	89.3	4,770	2,390	9,540	8,260	6,740	4,770	8,590	4,290	17,200	14,900	12,100	8,590
2	107.5	5,580	2,790	11,200	9,660	7,890	5,580	10,000	5,020	20,100	17,400	14,200	10,000
2 1/8	125.0	6,480	3,240	13,000	11,200	9,161	6,480	11,700	5,830	23,300	20,200	16,500	11,700
2 1/4	146.0	7,380	3,690	14,800	12,800	10,400	7,380	13,300	6,640	26,600	23,000	18,800	13,300
2 ½	166.7	8,370	4,190	16,700	14,500	11,800	8,370	15,100	7,530	30,100	26,100	21,300	15,100
2 5/8	190.8	9,360	4,680	18,700	16,200	13,200	9,360	16,800	8,420	33,700	29,200	23,800	16,800

See Figs. N-184-4 and N-184-5 for sling configuration descriptions.

Table 4.F.17. (N-184-17) Nylon Rope Slings

(Angle of rope to vertical shown in parentheses)

Rope dia. nominal in inches	Nominal wt. per 100 ft in pounds	Eye and eye sling						Endless sling					
		Vertical hitch	Choker hitch	Basket hitch; Angle of rope to horizontal				Vertical hitch	Choker hitch	Basket hitch; Angle of rope to horizontal			
				90° (0°)	60 (30°)	45 (45°)	30 (60°)			90° (0°)	60 (30°)	45 (45°)	30 (60°)
½	6.5	635	320	1,270	1,100	900	635	1,140	570	2,290	1,980	1,620	1,140
9/16	8.3	790	395	1,580	1,370	1,120	790	1,420	710	2,840	2,460	2,010	1,420
5/8	10.5	1,030	515	2,060	1,780	1,460	1,030	1,850	925	3,710	3,210	2,620	1,850
3/4	14.5	1,410	705	2,820	2,440	1,990	1,410	2,540	1,270	5,080	4,400	3,590	2,540
13/16	17.0	1,680	840	3,360	2,910	2,380	1,680	3,020	1,510	6,050	5,240	4,280	3,020
7/8	20.0	1,980	990	3,960	3,430	2,800	1,980	3,560	1,780	7,130	6,170	5,040	3,560
1	26.0	2,480	1,240	4,960	4,300	3,510	2,480	4,460	2,230	8,930	7,730	6,310	4,460
1 1/16	29.0	2,850	1,430	5,700	4,940	4,030	2,850	5,130	2,570	10,300	8,890	7,260	5,130
1 1/8	34.0	3,270	1,640	6,540	5,660	4,620	3,270	5,890	2,940	11,800	10,200	8,330	5,890
1 1/4	40.0	3,710	1,860	7,420	6,430	5,250	3,710	6,680	3,340	13,400	11,600	9,450	6,680
1 5/16	45.0	4,260	2,130	8,520	7,380	6,020	4,260	7,670	3,830	15,300	13,300	10,800	7,670
1 ½	55.0	5,250	2,630	10,500	9,090	7,420	5,250	9,450	4,730	18,900	16,400	13,400	9,450
1 5/8	68.0	6,440	3,220	12,900	11,200	9,110	6,440	11,600	5,800	23,200	20,100	16,400	11,600
1 3/4	83.0	7,720	3,860	15,400	13,400	10,900	7,720	13,900	6,950	27,800	24,100	19,700	13,900
2	95.0	9,110	4,560	18,200	15,800	12,900	9,110	16,400	8,200	32,800	28,400	23,200	16,400
2 1/8	109.0	10,500	5,250	21,000	18,200	14,800	10,500	18,900	9,450	37,800	32,700	26,700	18,900
2 1/4	129.0	12,400	6,200	24,800	21,500	17,500	12,400	22,300	11,200	44,600	38,700	31,600	22,300
2 ½	149.0	13,900	6,950	27,800	24,100	19,700	13,900	25,000	12,500	50,000	43,300	35,400	25,000
2 5/8	168.0	16,000	8,000	32,000	27,700	22,600	16,000	28,800	14,400	57,600	49,900	40,700	28,800

See Figs. N-184-4 and N-184-5 for sling configuration descriptions.

Table 4.F.18. (N-184-18) Polyester Rope Slings

(Angle of rope to vertical shown in parentheses)

Rope dia. nominal in inches	Nominal wt. per 100 ft in pounds	Eye and eye sling						Endless sling					
		Vertical hitch	Choker hitch	Basket hitch; Angle of rope to horizontal				Vertical hitch	Choker hitch	Basket hitch; Angle of rope to horizontal			
				90° (0°)	60 (30°)	45 (45°)	30 (60°)			90° (0°)	60 (30°)	45 (45°)	30 (60°)
½	8.0	635	320	1,270	1,100	900	635	1,140	570	2,290	1,980	1,620	1,140
9/16	10.2	790	395	1,580	1,370	1,120	790	1,420	710	2,840	2,460	2,010	1,420
5/8	13.0	990	495	1,980	1,710	1,400	990	1,780	890	3,570	3,090	2,520	1,780
3/4	17.5	1,240	620	2,480	2,150	1,750	1,240	2,230	1,120	4,470	3,870	3,160	2,230
13/16	21.0	1,540	770	3,080	2,670	2,180	1,540	2,770	1,390	5,540	4,800	3,920	2,770
7/8	25.0	1,780	890	3,560	3,080	2,520	1,780	3,200	1,600	6,410	5,550	4,530	3,200
1	30.5	2,180	1,090	4,360	3,780	3,080	2,180	3,920	1,960	7,850	6,800	5,550	3,920
1 1/16	34.5	2,530	1,270	5,060	4,380	3,580	2,530	4,550	2,280	9,110	7,990	6,440	4,550
1 1/8	40.0	2,920	1,460	5,840	5,060	4,130	2,920	5,260	2,630	10,500	9,100	7,440	5,260
1 1/4	46.3	3,290	1,650	6,580	5,700	4,650	3,290	5,920	2,960	11,800	10,300	8,380	5,920
1 5/16	52.5	3,710	1,860	7,420	6,430	5,250	3,710	6,680	3,340	13,400	11,600	9,450	6,680
1 ½	66.8	4,630	2,320	9,260	8,020	6,550	4,630	8,330	4,170	16,700	14,400	11,800	8,330
1 5/8	82.0	5,640	2,820	11,300	9,770	7,980	5,640	10,200	5,080	20,300	17,600	14,400	10,200
1 3/4	98.0	6,710	3,360	13,400	11,600	9,490	6,710	12,100	6,040	24,200	20,900	17,100	12,100
2	118.0	7,920	3,960	15,800	13,700	11,200	7,920	14,300	7,130	28,500	24,700	20,200	14,300
2 1/8	135.0	9,110	4,460	18,200	15,800	12,900	9,110	16,400	8,200	32,800	28,400	23,200	16,400
2 1/4	157.0	10,600	5,300	21,200	18,400	15,000	10,600	19,100	9,540	38,200	33,100	27,000	19,100
2 ½	181.0	12,100	6,050	24,200	21,000	17,100	12,100	21,800	10,900	43,600	37,700	30,800	21,800
2 5/8	205.0	13,600	6,800	27,200	23,600	19,200	13,600	24,500	12,200	49,000	42,400	34,600	24,500

See Figs. N-184-4 and N-184-5 for sling configuration descriptions.

Table 4.F.19. (N-184-19) Polypropylene Rope Slings

(Angle of rope to vertical shown in parentheses)

Rope dia. nominal in inches	Nominal wt. per 100 ft in pounds	Eye and eye sling Vertical hitch	Eye and eye sling Choker hitch	Basket hitch; Angle of rope to horizontal 90°(0°)	60(30°)	45(45°)	30(60°)	Vertical hitch	Choker hitch	Endless sling Basket hitch; Angle of rope to horizontal 90°(0°)	60(30°)	45(45°)	30(60°)
½	4.7	645	325	1,290	1,120	910	645	1,160	580	2,320	2,010	1,640	1,160
9/16	6.1	780	390	1,560	1,350	1,100	780	1,400	700	2,810	2,430	1,990	1,400
5/8	7.5	950	475	1,900	1,650	1,340	950	1,710	855	3,420	2,960	2,420	1,710
3/4	10.7	1,300	650	2,600	2,250	1,840	1,300	2,340	1,170	4,680	4,050	3,310	2,340
13/16	12.7	1,520	760	3,040	2,630	2,150	1,520	2,740	1,370	5,470	4,740	3,870	2,740
7/8	15.0	1,760	880	3,520	3,050	2,490	1,760	3,170	1,580	6,340	5,490	4,480	3,170
1	18.0	2,140	1,070	4,280	3,700	3,030	2,140	3,850	1,930	7,700	6,670	5,450	3,860
1 1/16	20.4	2,450	1,230	4,900	4,240	3,460	2,450	4,410	2,210	8,820	7,640	6,240	4,410
1 1/8	23.7	2,800	1,400	5,600	4,850	3,960	2,8000	5,040	2,520	10,100	8,730	7,130	5,040
1 1/4	27.0	3,210	1,610	6,420	5,560	4,540	3,210	5,780	2,890	11,600	10,000	8,170	5,780
1 5/16	30.5	3,600	1,800	7,200	6,240	5,090	3,600	6,480	3,240	13,000	11,200	9,170	6,480
1 ½	38.5	4,540	2,270	9,080	7,860	6,420	4,540	8,170	4,090	16,300	14,200	11,600	8,170
1 5/8	47.5	5,510	2,760	11,000	9,540	7,790	5,510	9,920	4,960	19,600	17,200	14,000	9,920
1 3/4	57.0	6,580	3,290	13,200	11,400	9,300	6,580	11,800	5,920	23,700	20,500	16,800	11,800
2	69.0	7,960	3,980	15,900	13,800	11,300	7,960	14,300	7,160	28,700	24,800	20,300	14,300
2 1/8	80.0	9,330	4,670	18,700	16,200	13,200	9,330	16,800	8,400	33,600	29,100	23,800	16,800
2 1/4	92.0	10,600	5,300	21,200	18,400	15,000	10,600	19,100	9,540	38,200	33,100	27,000	19,100
2 ½	107.0	12,200	6,100	24,400	21,100	17,300	12,200	22,000	11,000	43,900	38,100	31,100	22,000
2 5/8	120.0	13,800	6,900	27,600	23,900	19,600	13,800	24,800	12,400	49,700	43,000	35,100	24,800

See Figs. N-184-4 and N-184-5 for sling configuration descriptions.

(h)(3)

Splicing. Spliced fiber rope slings shall not be used unless they have been spliced in accordance with the following minimum requirements and in accordance with any additional recommendations of the manufacturer:

(h)(3)(i)

In manila rope, eye splices shall consist of at least three full tucks, and short splices shall consist of at least six full tucks, three on each side of the splice center line.

(h)(3)(ii)

In synthetic fiber rope, eye splices shall consist of at least four full tucks, and short splices shall consist of at least eight full tucks, four on each side of the center line.

(h)(3)(iii)

Strand end tails shall not be trimmed flush with the surface of the rope immediately adjacent to the full tucks. This applies to all types of fiber rope and both eye and short splices. For fiber rope under one inch in diameter, the tail shall project at least six rope diameters beyond the last full tuck. For fiber rope one inch in diameter and larger, the tail shall project at least six inches beyond the last full tuck. Where a projecting tail interferes with the use of the sling, the tail shall be tapered and spliced into the body of the rope using at least two additional tucks (which will require a tail length of approximately six rope diameters beyond the last full tuck).

(h)(3)(iv)

Fiber rope slings shall have a minimum clear length of rope between eye splices equal to 10 times the rope diameter.

(h)(3)(v)

Knots shall not be used in lieu of splices.

(h)(3)(vi)

Clamps not designed specifically for fiber ropes shall not be used for splicing.

(h)(3)(vii)

For all eye splices, the eye shall be of such size to provide an included angle of not greater than 60 degrees at the splice when the eye is placed over the load or support.

(h)(4)

End attachments. Fiber rope slings shall not be used if end attachments in contact with the rope have sharp edges or projections.

(h)(5)

Removal from service. Natural and synthetic fiber rope slings shall be immediately removed from service if any of the following conditions are present:

(h)(5)(i)

Abnormal wear.

(h)(5)(ii)

Powdered fiber between strands.

(h)(5)(iii)

Broken or cut fibers.

(h)(5)(iv)

Variations in the size or roundness of strands.

(h)(5)(v)

Discoloration or rotting.

(h)(5)(vi)

Distortion of hardware in the sling.

(h)(6) Repairs

Only fiber rope slings made from new rope shall be used. Use of repaired or reconditioned fiber rope slings is prohibited.

(i) Synthetic Web Slings

(i)(1)

Sling identification. Each sling shall be marked or coded to show the rated capacities for each type of hitch and type of synthetic web material.

(i)(2)

Webbing. Synthetic webbing shall be of uniform thickness and width and selvage edges shall not be split from the webbing's width.

(i)(3)

Fittings. Fittings shall be:

(i)(3)(i)

Of a minimum breaking strength equal to that of the sling; and

(i)(3)(ii)

Free of all sharp edges that could in any way damage the webbing.

(i)(4)

Attachment of end fittings to webbing and formation of eyes. Stitching shall be the only method used to attach end fittings to webbing and to form eyes. The thread shall be in an even pattern and contain a sufficient number of stitches to develop the full breaking strength of the sling.

(i)(5)

Sling use. Synthetic web slings illustrated in Fig. N-184-6 shall not be used with loads in excess of the rated capacities specified in Tables N-184-20 through N-184-22. Slings not included in these tables shall be used only in accordance with the manufacturer's recommendations.

(i)(6)

Environmental conditions. When synthetic web slings are used, the following precautions shall be taken:

(i)(6)(i)

Nylon web slings shall not be used where fumes, vapors, sprays, mists or liquids of acids or phenolics are present.

(i)(6)(ii)

Polyester and polypropylene web slings shall not be used where fumes, vapors, sprays, mists or liquids of caustics are present.

(i)(6)(iii)

Web slings with aluminum fittings shall not be used where fumes, vapors, sprays, mists or liquids of caustics are present.

See Figure 3.6 for basic synthetic web sling capacities.

(i)(7)

Safe operating temperatures. Synthetic web slings of polyester and nylon shall not be used at temperatures in excess of 180 deg. F. Polypropylene web slings shall not be used at temperatures in excess of 200 deg. F.

(i)(8) Repairs

(i)(8)(i)

Synthetic web slings which are repaired shall not be used unless repaired by a sling manufacturer or an equivalent entity.

(i)(8)(ii)

Each repaired sling shall be proof tested by the manufacturer or equivalent entity to twice the rated capacity prior to its return to service. The employer shall retain a certificate of the proof test and make it available for examination.

Table 4.F.20. (N-184-20) Synthetic Web Slings

1,000 Pounds Per Inch of Width -- Single-Ply
[Rated capacity in pounds]

Sling body width, inches	Triangle -- Choker slings, type I: Triangle -- Triangle slings, type II: Eye and eye with flat eye slings, type III: Eye and eye with twisted eye slings, type IV						Endless slings, type V						Return eye slings, type VI					
	Vert.	Choker	Vert. Basket	30° basket	45° basket	60° basket	Vert.	Choker	Vert. Basket	30° basket	45° basket	60° basket	Vert.	Choker	Vert. Basket	30° basket	60° basket	60° basket
1	1,000	750	2,000	1,700	81,400	91,000	1,600	1,300	3,200	2,800	2,300	1,600	800	650	1,600	1,400	1,150	800
2	2,000	1,500	4,000	3,500	82,800	92,000	3,200	2,600	6,400	5,500	4,500	3,200	1,600	1,300	3,200	2,800	2,300	1,600
3	3,000	2,200	6,000	5,200	84,200	93,000	4,800	3,800	9,600	8,300	6,800	4,800	2,400	1,950	4,800	4,150	3,400	2,400
4	4,000	3,000	8,000	6,900	85,700	94,000	6,400	5,100	12,800	11,100	9,000	6,400	3,200	2,600	6,400	5,500	4,500	3,200
5	5,000	3,700	10,000	8,700	87,100	95,000	8,000	6,400	16,000	13,900	11,300	8,000	4,000	3,250	8,000	6,900	5,650	4,000
6	6,000	4,500	12,000	10,400	88,500	96,000	9,600	7,700	19,200	16,600	13,600	9,600	4,800	3,800	9,600	8,300	6,800	4,800

Notes: 1. All angles shown are measured from the vertical.
2. Capacities for intermediate widths not shown may be obtained by interpolation.

Table 4.F.21. (N-184-21) Synthetic Web Slings

1,200 Pounds Per Inch of Width -- Single-Ply
[Rated capacity in pounds]

Sling body width, inches	Triangle -- Choker slings, type I: Triangle -- Triangle slings, type II: Eye and eye with flat eye slings, type III: Eye and eye with twisted eye slings, type IV						Endless slings, type V						Return eye slings, type VI					
	Vert.	Choker	Vert. Basket	30° basket	45° basket	60° basket	Vert.	Choker	Vert. Basket	30° basket	45° basket	60° basket	Vert.	Choker	Vert. Basket	30° basket	45° basket	60° basket
1	1,200	900	2,400	2,100	1,700	1,200	1,900	1,500	3,800	3,300	2,700	1,900	950	750	1,900	1,650	1,350	950
2	2,400	1,800	4,800	4,200	3,400	2,400	3,800	3,000	7,600	6,600	5,400	3,800	1,900	1,500	3,800	3,300	2,700	1,900
3	3,600	2,700	7,200	6,200	5,100	3,600	5,800	4,600	11,600	10,000	8,200	5,800	2,850	2,250	5,700	4,950	4,050	2,850
4	4,800	3,600	9,600	8,300	6,800	4,800	7,700	6,200	15,400	13,300	10,900	7,700	3,800	3,000	7,600	6,600	5,400	3,800
5	6,000	4,500	12,000	10,400	8,500	6,000	9,600	7,700	19,200	16,600	13,600	9,600	4,750	3,750	9,500	8,250	6,750	4,750
6	7,200	5,400	14,400	12,500	10,200	7,200	11,500	9,200	23,000	19,900	16,300	11,500	5,800	4,600	11,600	10,000	8,200	5,800

Notes: 1. All angles shown are measured from the vertical.
2. Capacities for intermediate widths not shown may be obtained by interpolation.

Table 4.F.22. (N-184-22) Synthetic Web Slings

1,600 Pounds Per Inch of Width -- Single-Ply
[Rated capacity in pounds]

Sling body width, inches	Triangle -- Choker slings, type I: Triangle -- Triangle slings, type II: Eye and eye with flat eye slings, type III: Eye and eye with twisted eye slings, type IV						Endless slings, type V						Return eye slings, type VI					
	Vert.	Choker	Vert. Basket	30° basket	45° basket	60° basket	Vert.	Choker	Vert. Basket	30° basket	45° basket	60° basket	Vert.	Choker	Vert. Basket	30° basket	45° basket	60° basket
1	1,600	1,200	3,200	2,800	2,300	1,600	2,600	2,100	5,200	4,500	3,700	2,600	1,050	1,050	2,600	2,250	1,850	1,300
2	3,200	2,400	6,400	5,500	4,500	3,200	5,100	4,100	10,200	8,800	7,200	5,100	2,600	2,100	5,200	4,500	3,700	2,600
3	4,800	3,600	9,600	8,300	6,800	4,800	7,700	6,200	15,400	13,300	10,900	7,700	3,900	3,150	7,800	6,750	5,500	3,900
4	6,400	4,800	12,800	11,100	9,000	6,400	10,100	8,200	20,400	17,700	14,400	10,200	5,100	4,100	10,200	8,800	7,200	5,100
5	8,000	6,000	16,000	13,800	11,300	8,000	12,800	10,200	25,600	22,200	18,100	12,800	6,400	5,150	12,800	11,050	9,050	6,400
6	9,600	7,200	19,200	16,600	13,600	9,600	15,400	12,300	30,800	26,700	21,800	15,400	7,700	6,200	15,400	13,300	1,900	7,700

Notes: 1. All angles shown are measured from the vertical.
  2. Capacities for intermediate widths not shown may be obtained by interpolation.

(i)(8)(iii)

Slings, including webbing and fittings, which have been repaired in a temporary manner shall not be used.

(i)(9) Removal from service.

Synthetic web slings shall be immediately removed from service if any of the following conditions are present:

(i)(9)(i)

Acid or caustic burns;

(i)(9)(ii)

Melting or charring of any part of the sling surface;

(i)(9)(iii)

Snags, punctures, tears or cuts;

(i)(9)(iv)

Broken or worn stitches; or

(i)(9)(v)

Distortion of fittings.

[40 FR 27369, June 27, 1975, as amended at 40 FR 31598, July 28, 1975; 41 FR 13353, Mar. 30, 1976; 58 FR 35309, June 30, 1993; 61 FR 9227, March 7, 1996]

REFERENCES

1. Occupational Safety and Health Administration. OSHA, 1910.179 — Overhead and gantry cranes.

2. Occupational Safety and Health Administration. OSHA, 1910.180 — Crawler locomotive and truck cranes.

3. Occupational Safety and Health Administration. OSHA, 1910.181 — Derricks.

4. Occupational Safety and Health Administration. OSHA, 1910.184 — Slings.

5. Occupational Safety and Health Administration. OSHA, 1926.550 — Cranes and derricks.

6. Occupational Safety and Health Administration. OSHA, 1926.251 — Rigging equipment for material handling.

7. The Crosby Group, Inc.. *Crosby User's Lifting Guide*, Tulsa, OK. 1998, pp. 1-12.

5

TRAINING REQUIREMENTS

Training and qualification requirements are specified by applicable OSHA and ANSI standards, rules, and regulations for personnel who direct or perform hoisting and rigging work. OSHA specifies that the employer shall permit only those qualified by training or experience to operate equipment. Equipment in this instance refers to cranes and derricks. ANSI states that it is essential to have personnel involved in the use of equipment who are competent, careful, physically and mentally qualified, and trained in the safe operation of the equipment and the handling of the loads. All personnel who are designated to operate equipment or perform work covered by these requirements must be trained and qualified to the level of proficiency consistent with assigned tasks. This includes supervisors (or any personnel involved in planning work), operators, riggers, inspectors, and training instructors. If an individual other than those listed here is used to signal the crane operator, then that individual is also required to be trained for their specific task. Ineffective training or the lack of hoisting and rigging training can lead to personnel events ranging from minor injuries to fatalities as well as the loss of the hoisting and rigging equipment, material being lifted, and other property in the work zone.

Accidents while using cranes to hoist material occur frequently costing lives, injuries, and property damage. These accidents are not unique. Most of them have the same causes: equipment failure or employee error. Serious hazards are overloading, dropping or slipping of the load caused by using improper hitches or slings, obstructing the free passage of the load, and using equipment for a purpose for which it was not intended or designed. Equipment failure is often due to improper use, not knowing the weight of the load, or failing to inspect the equipment before use. Employee error is often caused by inattention to safe work practices or by not having been trained on the correct use of the equipment. To safely operate any crane requires an understanding of the operating principles and procedures for the specific equipment. Employees must understand that they are to read and heed all warning tags and signs. This includes items such as staying clear of pinch points on cranes and of suspended loads.

According to the Department of Labor, 15 crane operators experienced fatal occupational injuries in 1997. While some were not performing hoisting and rigging work at the time, another 138 employees died in 1997 after coming in contact with overhead power lines. Employees must be trained to notify their immediate supervisor if they are assigned tasks that they are not qualified to perform. Employers are responsible for ensuring that work assignments do not exceed personnel qualifications, and employers maintaining a list of qualified operators on appropriate equipment is recommended as one way of ensuring this. An operator is defined by ANSI as an individual specifically designated by the employer to operate a specified crane.

Training Providers

The following criteria can be used in the selection of a training organization to provide training, evaluations, and qualifications to meet OSHA and ANSI requirements:

- Review available training classes with subject matter experts

- Draft statement of work with specific requirements

- Obtain and evaluate the following from a prospective training provider:
  - All written materials provided in training
  - Copy of all classroom and field exercises
  - All instructional materials used in training
  - Lesson plan(s) guide, course outline with learning objectives, transparencies, slides, and equipment/tools as needed (some companies may be reluctant to provide some or all of the material mentioned)
  - Course examinations, evaluations, job performance measurements with answer keys
  - A description of facilities and equipment used to support the course
  - Location of training
  - Cost
  - Instructor(s) qualifications.

Training programs that are used to meet qualification requirements must include clear objectives, with elements as applicable, that require the students to:

- Identify equipment components

- Perform pre-use inspection

- Verify load capacity of equipment/cranes for given configuration

- Define roles and responsibilities

- Demonstrate safe operating practices

- Identify potential hazards and controls

- Demonstrate proper equipment setup/storage

- Calculate load weights

- Demonstrate knowledge of applicable codes, standards, and requirements for equipment and personnel

- Demonstrate skills in operation and use of slings, rigging hardware, below-the-hook lifting devices, and cranes as applicable

- Perform periodic documented inspections (qualified inspectors only)

- Operate cranes that they will be authorized to operate as part of an on-the-job evaluation.

Physical Requirements For Employees

In addition to the requirement that employees be trained, those who work within the scope of these standards are also required to be able to meet all physical requirements for their assignment. As mentioned earlier, prior to operating mobile, locomotive, and cab- or pulpit-operated overhead cranes, operators and operator trainees are required to pass a physical exam which must be in accordance with consensus standards such as ASME B30.2 (Overhead and Gantry Cranes) and ASME B30.5 (Mobile and Locomotive Cranes). Operator physical examinations should be required every three years, or more frequently if an employer deems it necessary. Exceptions, deviations, or additions to consensus standard requirements for physical requirements must be approved by OSHA.

Drug Testing/Drivers License

The level of drug testing is determined by the standard practice for the industry where the crane is operated and this test shall be confirmed by a recognized laboratory service. In all instances, operators and operators-in-training of mobile or other cranes must be required to pass with a negative result a substance abuse test prior to initial employment and substance abuse retesting on a periodic basis or random testing at the employer's discretion. This also includes any maintenance personnel that drive mobile cranes on public roads. The Commercial Motor Vehicle Safety Act of 1986 requires drivers of vehicles with a gross or registered weight in excess of 26,000 pounds to have a commercial drivers license (CDL). Consequently, this Act applies to mobile cranes within the scope of these standards driven on public roads.

Required Training for Employees

Training, evaluations, and on-the-job training (OJT) is required by the respective OSHA and ANSI standards for personnel who direct, operate, or perform work for the following activities:

- Rigging
- Equipment inspection
- Mobile crane operation
- Overhead crane operation.

ANSI requires employers to permit only those employees who have passed a written or oral examination and a practical operation examination to perform the activities or operate the equipment referenced above. How employers provide the training in meeting those requirements remains the employer's option. Training may include, but is not limited to, computer-aided training, classroom training, simulated field training, and on-the-job training. The training may offer varying levels of knowledge and skills appropriate to the equipment and complexity of work and include applicable rules, standards, and regulations. Any qualification that has been verified by a physical, written, and practical examination and provided to the employee is limited to the specific crane for which the employee was trained.

A qualified rigger is an individual with rigging experience and has demonstrated competence as a rigger to a person with authority and having jurisdiction to make the determination that the person is qualified. Some of the issues the individual needs to have a working knowledge of include knowing how to:

- Determine the weight of a load
- Locate the center of gravity of a load
- Choose the correct rigging hardware (shackles, hooks, eyebolts, etc.)
- Identify the different wire rope components
- Identify factors affecting wire rope strength
- Determine wire rope, chain, and web sling capacities

- Determine the limitation of the system chosen to make a lift

- Figure the load on legs of a sling at different angles

- Inspect slings (wire rope, chain, and web) and remove from service when warranted due to the following conditions:

  - Kinking
  - Birdcaging
  - Crushing
  - Broken strands
  - Unstranding
  - Heat damage
  - Outer strands distortion
  - Main strand displacement
  - Core protrusion
  - Reduced rope diameter
  - Rust or corrosion.

Crane inspections can be performed by the operator or other employer-designated personnel. "Designated" means selected or assigned by the employer or employer's representative as being qualified to perform specific duties. The employee chosen to inspect cranes must be trained in the requirements of inspection classifications; initial, frequent, and periodic. To perform the initial inspection of new, reinstalled, altered, repaired, and modified cranes, the operator or other designated person is required to be trained in performing inspections to verify compliance with all applicable regulations.

For frequent and periodic inspections, the operator or designated person is required to know the two general classifications (frequent and periodic inspection) at which inspections are to be performed as well as the conditions to be observed during the inspection. Frequent inspections require visual examinations (with no records required to be kept) at three different intervals: (1) monthly for normal service, (2) weekly to monthly for heavy service, and (3) daily to weekly for severe service. Periodic inspections of cranes require a visual inspection the same as frequent inspections, except the designated person is required to make a record of any apparent external conditions as a basis to provide a continuing evaluation for the following intervals (1) yearly for normal or heavy service, and (2) quarterly for severe service.

Employers of crane operators are required to conduct the equipment-specific familiarization and evaluation program applicable to the type of crane demonstrating configurations, equipment applications, and organizational specific rules. An employee's documented evidence of previous training and experience may be accepted by an employer. When previous training is accepted, documented written and performance evaluations must be required by the employer before qualification is granted and before an employee is allowed to operate a crane. To be acceptable to an employer for qualification, there must be documented evidence indicating type and class of equipment and hours of experience and training authenticated by the responsible organization. Examples of previous training may include:

- Equipment manufacturer or distributor training

- Completion of an applicable apprenticeship program

- Journeyman status in an applicable trade.

Training Documentation and Evaluation

All required training must be documented, including written tests and performance evaluations. These tests and evaluations are used for the purpose of requiring personnel to demonstrate the knowledge and skill to safely perform the following elements of hoisting and rigging operations as applicable to the type of crane they will be operating:

- Pre-use inspection

- Load rating charts

- Controls/function/position/limits

- Operating characteristics

- Emergency operating procedures

- Response to fire

- Working near and response to power line contact

- Equipment set-up

- Shut down and park

- Attachments

- Equipment configurations/loading/effects

- Current applicable standards/rules/regulations

- Rigging practices

- Sling loading and effects on sling capacities

- Weight calculations

- Designated Leader responsibilities.

Employers are required to requalify employees at specific intervals after the successful completion of initial qualification requirements. Employees must be required to requalify three years from date of initial training, evaluation, or completion of OJT. Any employees with expired qualifications are considered to be unqualified. Qualification records must be maintained by the employer for the duration of qualification. The employee record must contain the written results and performance evaluation of knowledge and skills. To requalify, employees have the option of one of the following:

- Successful completion of a written and/or performance challenge evaluation
- Completion of training and evaluations.

Crane Accidents

A review of accidents shows that the single greatest contributor is most often human error, caused by the lack of, or improper, training. One common type of accident is overturned cranes, with some of the causes being:

- Improper crane setup (i.e., outriggers not used, crane not set up level)
- Crane's rated capacity exceeded
- Maximum operating radius improperly determined
- Structural chart capacity exceeded
- Errors by team members
- Load weight not correctly determined
- Wrong slings and attachments (type, size, etc.) used
- Slings improperly attached to the load
- Slings and/or hardware not inspected
- Center of gravity not calculated
- Electrical power lines not taken into consideration
- Weather conditions not taken into consideration:
  - Wind speed
  - Lightning
  - Fog
  - Ice
  - Visibility (darkness)
  - Rain (soft ground).

Questions on Mobile Crane Safety

T = True, F = False   *T* *F*

1. The lower load block and hook is not considered part of the load.   ☐ ☐

2. Load ratings for some radii are limited by the stability of the crane.   ☐ ☐

3. Durable rating charts are required only on some cranes.   ☐ ☐

4. Boom is always positioned at minimum recommended boom angle.   ☐ ☐

5. No less than two full wraps of rope shall remain on the drum when the hook
    is in the extreme low position.   ☐ ☐

6. Extension and retraction of boom sections may only be accomplished mechanically. ☐ ☐

7. An angle indicator measures the angle of the boom to the vertical.   ☐ ☐

8. The boom hoist is a hoist drum and rope reeving system used to raise and lower
    the boom.   ☐ ☐

9. A jib is an extension attached to the boom point to provide added boom length for
    lifting specified loads.   ☐ ☐

10. Load rating are established by the crane owner.   ☐ ☐

11. Outriggers are extendible or fixed metal arms used when the operators decides to. ☐ ☐

12. Reeving means a rope system in which the rope travels around drums and sheaves. ☐ ☐

13. Rope refers to a wire rope unless otherwise specified.   ☐ ☐

14. Swing means the amount of boom sway in high wind.   ☐ ☐

15. Tackle is an assembly of ropes and sheaves arranged for hoisting and pulling. ☐ ☐

16. Wheelbase means the length of the axles.   ☐ ☐

17. Designated personnel can authorize employees as crane operators.   ☐ ☐

18. New and altered cranes do not require an initial inspection.   ☐ ☐

19. Frequent inspections are done at daily to monthly intervals.   ☐ ☐

20. Periodic inspections are performed when the operator has time.   ☐ ☐

21. Standby cranes need no inspection until put in service.   ☐ ☐

22. Test loads shall not exceed 125 percent of the rated load.   ☐ ☐

23. Running ropes require inspection at least annually. ☐ ☐

24. The hoist rope shall not be wrapped around the load. ☐ ☐

25. Cranes shall not be used for dragging loads sideways. ☐ ☐

26. "ANSI" means the American National Standards Institute ☐ ☐

27. Only appointed personnel shall be permitted to operate a crane. ☐ ☐

28. A load rating chart must be visible to all employees on the job site. ☐ ☐

29. New cranes need no inspection prior to initial use. ☐ ☐

30. Frequent inspections are performed at daily to monthly intervals. ☐ ☐

31. A thorough inspection of running ropes is required daily. ☐ ☐

32. It is permissible to load the crane beyond the rated load as necessary. ☐ ☐

33. The hoist rope shall not be wrapped around the load. ☐ ☐

34. The operator should avoid carrying loads over people. ☐ ☐

35. In transit, the boom shall be carried in line with the direction of motion. ☐ ☐

36. The empty hook can swing freely in transit. ☐ ☐

37. The operator is permitted to leave his position at the controls while a load is suspended. ☐ ☐

38. Persons are permitted to stand under a load only for a short time period. ☐ ☐

39. Tools are permitted to lie loose in or about the cab. ☐ ☐

40. Operations near overhead lines require no special considerations. ☐ ☐

Answers to Questions on Mobile Crane Safety

1.	False	22.	False
2.	True	23.	False
3.	False	24.	True
4.	False	25.	True
5.	True	26.	True
6.	False	27.	False
7.	False	28.	False
8.	True	29.	False
9.	True	30.	True
10.	False	31.	False
11.	False	32.	False
12.	True	33.	True
13.	True	34.	True
14.	False	35.	True
15.	True	36.	False
16.	False	37.	True
17.	False	38.	False
18.	False	39.	False
19.	True	40.	False
20.	False		
21.	False		

Questions on Overhead and Gantry Crane Safety

T = True, F = False

		T	F
1.	A "holding brake" automatically prevents motion when power is off.	☐	☐
2.	"Exposed" means capable of being contacted inadvertently.	☐	☐
3.	Cranes may be modified and re-rated by the crane operator.	☐	☐
4.	The rated load marking shall be plainly marked on the cab of the crane.	☐	☐
5.	Light in the cab shall be sufficient to enable the operator to see clearly enough to perform his work.	☐	☐
6.	Footwalks require no less than 24 inches of provided headroom.	☐	☐
7.	Temporary ladders are permitted to be used as necessary.	☐	☐
8.	Stops shall be provided at the limits of travel of the trolley.	☐	☐
9.	Bridge bumpers shall be so mounted that there is no direct shear on bolts.	☐	☐
10.	Guards are never needed for hoisting ropes.	☐	☐
11.	Exposed moving parts which might constitute a hazard under normal operating conditions shall be guarded.	☐	☐
12.	Independent hoisting units of a crane require no holding brake.	☐	☐
13.	Holding brakes on hoists shall be applied automatically when power is removed.	☐	☐
14.	The wearing surface of all holding-brake drums or discs shall be smooth.	☐	☐
15.	A bridge brake is required on cab-operated cranes with cab on trolley.	☐	☐
16.	The control circuit voltage shall not exceed 400 volts for ac or dc current.	☐	☐
17.	The voltage at pendant push-buttons shall not exceed 120 volts for ac.	☐	☐
18.	Pendant control boxes shall be clearly marked for identification of functions.	☐	☐
19.	Live parts of electrical equipment shall be so located as to not be exposed to accidental contact under normal operating conditions.	☐	☐
20.	Electric equipment shall be protected from dirt, grease, oil, and moisture.	☐	☐

21. Pushbuttons in pendant stations shall be manually turned to the "off" position when released by the operator. ☐ ☐

22. A nonconductive rope attached to the main disconnect switch is a type of floor-operated disconnect switch. ☐ ☐

23. A service receptacle provided in the cab of a cab-operated crane shall not exceed 440 volts. ☐ ☐

24. Means for discharging the inductive load of a lifting magnet shall be provided. ☐ ☐

25. Sheave grooves should be smooth and free from surface defects. ☐ ☐

26. All running sheaves shall be equipped with means for lubrication. ☐ ☐

27. In hoisting ropes, the crane manufacturer's recommendations shall be used. ☐ ☐

28. No less than three wraps of rope shall remain on the drum when the hook is in the extreme low position. ☐ ☐

29. Rope ends anchored to the drum need not necessarily be approved by the crane manufacturer. ☐ ☐

30. Rope clips attached with U-bolts shall have the U-bolts on the dead or short end of the rope. ☐ ☐

31. Swaged or compressed fittings may be applied as deemed necessary. ☐ ☐

32. Replacement rope must be larger than the original rope. ☐ ☐

33. If a load is supported by more than one part of a rope, the tension in the parts shall be equalized. ☐ ☐

34. Hooks shall meet the manufacturer's recommendations and shall not be overloaded. ☐ ☐

35. No warning devices are required on overhead cranes. ☐ ☐

36. Frequent inspections include inspecting hooks for deformation or cracks. ☐ ☐

37. An electronic system that includes all of the required information would meet the intent of the recordkeeping requirements. ☐ ☐

38. There is no requirement that a preventive maintenance program be established. ☐ ☐

39. A thorough inspection of all running ropes is required at least annually. ☐ ☐

40. Fire extinguishers are required in all overhead crane cabs. ☐ ☐

GLOSSARY

The following are specialized terms commonly used when discussing hoisting and rigging operations. Some may not be used in this text but are included for general information. The terms are arranged in alphabetical order.

A

ABRASION: Frictional surface wear on the wires of a wire rope.

ACCESSORY: A secondary part or assembly of parts which contributes to the overall function and usefulness of a machine.

ACCELERATION STRESS: Additional stress that is imposed on a wire rope due to increase in the load velocity.

ALTERNATE LAY: Lay of wire rope in which the strands are alternately regular and lang lay.

ANSI: American National Standards Institute.

ANGLE INDICATOR [Boom]: An accessory that measures the angle of the boom to the horizontal.

ANGLE OF LOADING: The inclination of a leg or branch of a sling measured from the horizontal or vertical plane provided that an angle of loading of five degrees or less from the vertical may be considered a vertical angle of loading.

APPOINTED: Assigned specific responsibilities by the employer or the employer's representative.

AREA, METALLIC: Sum of the cross-sectional areas of individual wires in a wire rope or strand.

ASME: American Society of Mechanical Engineers.

AUTHORIZED: Assigned by a duly constituted administrative or regulatory authority.

AUXILIARY HOIST: Supplemental hoisting unit of lighter capacity and usually higher speed than the main hoist.

AXIS OF ROTATION: The vertical axis around which the crane superstructure rotates.

AXLE: The shaft or spindle with which or about which a wheel rotates. On truck- and wheel-mounted cranes it refers to an automotive type of axle assembly including housings, gearing, differential, bearings, and mounting appurtenances.

AXLE (Bogie): Means two or more automotive-type axles mounted in tandem in a frame so as to divide the load between the axles and permit vertical oscillation of the wheels.

B

BACK STAY: Guy used to support a boom or mast or that section of a main cable, as on a suspension bridge, or cableway, and the like, leading from the tower to the anchorage.

BAIL: A U-shaped member of a bucket, socket, or other fitting used on wire rope.

BASKET OR SOCKET: The conical portion of a socket into which a splayed rope end is inserted and secured with zinc.

BASKET HITCH: A sling configuration whereby the sling is passed under the load and has both ends, end attachments, eyes, or handles on the hook or a single master link.

BASE MOUNTING: The traveling base or carrier on which the rotating superstructure is mounted, such as a car, truck, crawlers, or wheel platform.

BECKET: An end attachment to facilitate wire rope attachment.

BECKET LOOP: A loop of small rope or a strand of rope fastened to the end of a large wire rope to facilitate wire rope installation.

BENDING STRESS: Stress on wires of a wire rope imposed by bending or curving action. This stress need not be added to direct load stresses. When sheaves and drums are of suitable size, bending stress does not affect the normal life of the wire rope.

BIRDCAGE: A colloquialism describing the appearance of a wire rope that is forced into compression. The outer strands form a "cage" and at times displace the core.

BIRDCAGING: The twisting of fiber or wire rope in an isolated area in the opposite direction of the rope lay, causing it to take on the appearance of a birdcage.

BOOM ANGLE: The angle between the longitudinal centerline of the boom and the horizontal. The boom longitudinal centerline is a straight line between the boom foot pin (heel pin) centerline and boom point sheave pin centerline.

BOOM (Crane): A member hinged to the front of the rotating superstructure with the outer end supported by ropes leading to a gantry or A-frame and used for supporting the hoisting tackle.

BOOM HOIST: A hoist drum and rope reeving system used to raise and lower the boom. The rope system may be all live reeving or a combination of live reeving and pendants.

BOOM HOIST LINE: A wire rope for supporting or operating the boom on derricks, cranes, draglines, shovels, and the like.

BOOM STOP: A device used to limit the angle of the boom at the highest position.

BRAIDED WIRE ROPE: A wire rope formed by plaiting component wire ropes.

BRAKE: A device used for retarding or stopping motion by friction or power means.

BRAKE, DRAG: A brake that provides stopping force without external control.

BRAKE, HOLDING: A brake that sets automatically and that prevents motion when power is off.

BRAKE, PARKING: A device to prevent the movement of a stationary vehicle.

BRAKING, COUNTERTORQUE: A method of stopping motion in which the power to the motor is reversed to develop torque in the opposite direction.

BRAKING, DYNAMIC: A method of controlling crane motor speeds when in the overhauling condition to provide a retarding force.

BRAKING, MECHANICAL: A method of slowing motion by friction.

BRAKING, REGENERATIVE: A form of dynamic braking in which the electrical energy generated is fed back into the power system.

BREAKING STRENGTH: The measured (ultimate) load required to break a wire rope or chain.

BRIDGE: The part of a crane, consisting of girders, walkways, railings, trucks, and drive mechanisms, that carries the trolley or trolleys.

BRIDGE TRAVEL: Horizontal travel of the crane parallel with runway rails.

BRIDLE SLING: A sling composed of multiple legs (branches), the top ends of which terminate in a fitting that latches onto the lifting hook.

BRIDLE WIRE ROPE SLING: A sling composed of multiple wire rope legs with the top ends gathered in a fitting that goes over the lifting hook.

BULL RING: The main large ring of a sling to which sling legs are attached.

BUMPER (Buffer): An energy-absorbing device for reducing impact when a moving overhead crane or trolley reaches the end of its permitted travel, or when two moving cranes or trolleys come into contact.

C

CAB: The operator's compartment on a crane. Housing which covers the rotating superstructure machinery and/or operator's station. On truck-crane trucks a separate cab covers the driver's station.

CABLE: A term loosely applied to wire ropes, wire strands, manila ropes, and electrical conductors.

CABLE LAID ENDLESS SLING-MECHANICAL JOINT: A wire rope sling made endless by joining the ends of a single length of cable laid rope with one or more metallic fittings.

CABLE LAID GROMMET-HAND TUCKED: An endless wire rope sling made from one length of rope wrapped six times around a core formed by hand tucking the ends of the rope inside the six wraps.

CABLE LAID ROPE: A wire rope composed of six wire ropes wrapped around a fiber or wire rope core.

CABLE LAID ROPE SLING-MECHANICAL JOINT: A wire rope sling made from a cable laid rope with eyes fabricated by pressing or swaging one or more metal sleeves over the rope junction.

CABLE-LAID WIRE ROPE: A type of wire rope consisting of several independent wire ropes laid into a single wire rope.

CABLE CROWD ROPE: A wire rope used to force the bucket of a power shovel into the material being handled.

CARRIAGE: A support structure for forks or attachments, generally roller-mounted, traveling vertically within the mast of a cantilever truck.

CENTER: The axial member (a single wire or fiber) in the center of a strand around which the wires are laid.

CENTER CONTROL: The position near the center of a truck cab from which the operator controls movement of the truck.

CHOKER HITCH: A sling configuration with one end of the sling passing under the load and through an end attachment, handle or eye on the other end of the sling.

CHOKER ROPE: A short wire-rope sling used to form a slip noose around the object to be moved or lifted.

CIRCUMFERENCE: Measured perimeter of a circle circumscribing the wires of a strand or the strands of a wire rope.

CLAMP, STRAND: A fitting used to form a loop at the end of a length of strand; consists of two grooved plates and bolts.

CLEARANCE: The distance by which one object clears another, or the clear space between them (i.e., the distance between the power lines and any part of the crane or its load).

CLEVIS: A U-shaped fitting with pins.

CLIP: A fitting used to clamp two parts of wire rope.

CLOSED SOCKET: A wire-rope fitting consisting of an integral becket and bail.

CLOSING LINE: Wire rope that closes a clamshell or orange-peel bucket and then operates as a hoisting rope.

CLUTCH: A friction, electromagnetic, hydraulic, pneumatic, or positive mechanical device for engagement or disengagement of power.

COATING: An elastomer or other suitable material applied to a sling or to a sling component to impart desirable properties.

COIL: Circular bundle of wire rope not packed on a reel.

COLLECTORS CURRENT: Contacting device mounted on a bridge or trolley and used to collect current from the conductor system.

CONDUCTOR: Wire, angles, bars, tees, or special sections mounted to transmit current to the collectors.

CONICAL DRUM: Grooved hoisting drum of varying diameter.

CONSTRUCTION (WIRE ROPE): Refers to the design of wire rope, including number of strands, number of wires per strand, and arrangement of wires in each strand.

CONTINUOUS BEND: Reeving of wire rope over sheaves and drums so that it bends in one direction (as opposed to reverse bend).

CONTROLLER: An operator's device for regulating the power delivered to a motor or other equipment.

CONTROLLER, SPRING RETURN: A controller that, when released, will return automatically to a neutral position.

CORE: The center member of a wire rope around which the strands are laid. It may be fiber, a wire strand, or an independent wire rope.

CORROSION: Chemical decomposition by exposure to moisture, acids, alkalies, or other destructive agents.

CORRUGATED: A term used to describe the grooves of a sheave or drum when worn so as to show the impression of a wire rope.

COUNTERTORQUE: A method of control by which the power to the crane motor is reversed to develop torque in the opposite direction.

COUNTERWEIGHT: A weight used to supplement the weight of the machine in providing stability for lifting working loads.

COVER WIRES: The outer layer of wires.

CRANE: A machine used for lifting and lowering a load vertically and moving it horizontally, with the hoisting mechanism an integral part of the machine. Cranes whether fixed or mobile are driven manually or by power.

CRANES, TYPES OF:

Automatic Crane: A crane that, when activated, operates through a preset cycle or cycles.

Cab-Operated Crane: A crane controlled by an operator in a cab located on the bridge or trolley.

Cantilever Gantry Crane: A gantry or semigantry crane in which the bridge girders or trusses extend transversely beyond the crane runway on one or both sides.

Crawler Crane: Consists of a rotating superstructure with power plant, operating machinery, and boom, mounted on a base, equipped with crawler treads for travel. Its function is to hoist and swing loads at various radii.

Floor-Operated Crane: A crane whose operation is controlled by use of a pendant in the hands of an operator on the floor or on an independent platform.

Gantry Crane: A crane similar to an overhead crane, except that the bridge for carrying the trolley or trolleys is rigidly supported on two or more legs running on fixed rails or another runway.

Hot Metal Handling Crane: An overhead crane used for transporting or pouring molten material.

Jib Crane: A fixed crane with a vertical rotating member supported at the bottom (also at the top in some types) from which an arm extends to carry the hoist trolley. Jib cranes are most commonly mounted on a vertical column, supplied as part of the jib crane, or on existing structural members (e.g., a wall-mounted jib crane).

Locomotive Crane: Consists of a rotating superstructure with power-plant, operating machinery, and boom, mounted on a base or car equipped for travel on railroad track. It may be self-propelled or propelled by an outside source. Its function is to hoist and swing loads at various radii.

Overhead Traveling Crane: A crane with a movable bridge carrying a movable or fixed hoisting mechanism and traveling on an overhead fixed-runway structure.

Power-Operated Crane: A crane whose mechanism is driven by electricity, air, hydraulics, or internal combustion.

Pulpit-Operated Crane: A crane operated from a fixed operator station that is not attached to the crane.

Remote-Operated Crane: A crane controlled by an operator not in a pulpit or a cab attached to the crane, by any method other than pendant or rope control (e.g., radio-controlled crane).

Semigantry Crane: A gantry crane with one end of the bridge rigidly supported on one or more legs that run on a fixed rail or runway, the other end of the bridge being supported by a truck running on an elevated rail or runway.

Storage Bridge Crane: A gantry type crane of long span usually used for bulk storage of material; the bridge girders or trusses are rigidly or non-rigidly supported on one or more legs. It may have one or more fixed or hinged cantilever ends.

Tower Crane: Consists of a rotating superstructure, made up of operating machinery, cab and boom, that rests on a circular track atop a tower.

Truck Crane: Consists of a rotating superstructure with powerplant, operating machinery and boom, mounted on an automotive truck equipped with a powerplant for travel. Its function is to hoist and swing loads at various radii.

Wall-Mounted Crane: A crane having a jib, with or without a trolley, supported from a side wall or line of columns of a building. It is a traveling-type crane and operates on a runway attached to the side wall or line of columns.

Wheel Mounted Crane: Consists of a rotating superstructure with powerplant, operating machinery and boom, mounted on a base or platform equipped with axles and rubber-tired wheels for travel. The base is usually propelled by the engine in the superstructure, but it may be equipped with a separate engine controlled from the superstructure. Its function is to hoist and swing loads at various radii.

CRITICAL DIAMETER: Diameter of the smallest bend for a given wire rope that permits the wires and strands to adjust themselves by relative movement while remaining in their normal positions.

CROSS ROD: A wire used to join spirals of metal mesh to form a complete fabric.

CYLINDRICAL DRUM: Hoisting drum of uniform diameter.

D

DECELERATION STRESS: Additional stress imposed on a wire rope due to decreasing the load velocity.

DEFLECTION:

▸ Sag of a rope in a span, usually measured at midspan as the depth from a chord joining the tops of the two supports.

▸ Any deviation from a straight line.

DERRICK: A mast or equivalent member held by guys or braces, with or without a boom, for use with a hoisting mechanism and operating ropes.

DESIGN (SAFETY) FACTOR: An industry term denoting a product's theoretical reserve capability; usually computed by dividing the ultimate load by the Working Load Limit. A ratio of ultimate strength to the design working stress; generally expressed as a ratio (e.g., 5 to 1).

DESIGNATED: Selected or assigned by the employer or the employer's representative as being qualified to perform specific duties.

DIAMETER: Distance measured across the center of a circle circumscribing the wires of a strand or the strands of a wire rope.

DOGLEG: Permanent short bend or kink in a wire rope caused by improper use or handling. This occurs when a wire rope sling is pulled down snug against a load. A dogleg usually can be "rolled back" or turned inside out, and usefulness of the sling restored, since strands can still adjust.

DRAGLINE: Wire rope used to pull an excavating or drag bucket.

DRIFT POINT: A point on a travel motion controller which releases the brake while the motor is not energized. This allows for coasting before the brake is set.

DRIVE: Motor, coupling, brake, and gear case or gear cases used to propel bridge, trolley, or hoist.

DRIVE GIRDER: A girder on which is mounted the bridge drive, cross shaft, walk, railing, and operator's cab.

DYNAMIC (Loading): Loads introduced into the machine or its components by forces in motion. A method of controlling crane motor speeds when in the overhauling condition to provide a retarding force.

DRUM: A cylindrical-flanged barrel of uniform (cylindrical drum) or tapering (conical drum) diameter on which a wire rope is wound for raising or lowering the load.

E

ELASTIC LIMIT: Limit of stress beyond which a permanent deformation takes place within the material. This limit is approximately 55–65 percent of breaking strength of steel-wire ropes.

EQUIVALENT ENTITY: A person or organization (including an employer) that, by possession of equipment, technical knowledge and skills, can perform with equal competence the same repairs and tests as the person or organization with which it is equated.

EQUALIZER: A device used to compensate for unequal length or stretch of a hoist rope.

EQUALIZING SLINGS: Slings composed of wire rope and equalizing fittings.

EQUALIZING THIMBLES: A special type of fitting used as a component part of some wire-rope slings.

EXPOSED: Capable of being contacted inadvertently. Applied to hazardous objects not adequately guarded or isolated.

EYE OR EYE SPLICE: A loop with or without a thimble formed in the end of a wire rope.

F

FABRIC (metal mesh): The flexible portion of a metal mesh sling consisting of a series of transverse coils and cross rods.

FAIL-SAFE: A provision designed to automatically stop or safely control any motion in which a malfunction occurs.

FAILURE: Means load refusal, breakage, or separation of components.

FATIGUE: A term commonly applied to progressive fracture of any load-supporting member.

FEMALE HANDLE (choker): A handle with a handle eye and a slot of such dimension as to permit passage of a male handle, thereby allowing the use of a metal mesh sling in a choker hitch.

FIBER CENTER: Cords or rope made of vegetable fiber used in the center of a strand.

FIBER CORE: Cords or rope made of vegetable fiber used in the core of a wire rope.

FIRST POINT: The first setting on the operator's controller that starts crane motion (slowly) in each direction.

FITTING: Any accessory used as an attachment for wire rope.

FLAG: Mark or marker on a rope to designate position of load.

FLAT ROPE: Wire rope made of parallel alternating right-lay and left-lay ropes sewn together by relatively soft wires.

FLATTENED STRAND ROPE: A wire rope with either oval or triangular strands that present a flattened rope surface.

FLEET ANGLE: Angle between the position of a rope at the extreme end wrap on a drum and a line drawn perpendicular to the axis of the drum through the center of the nearest fixed sheave.

FOOTWALK: The walkway with handrail, attached to the crane bridge or trolley for access purposes.

G

GALVANIZE: To coat with zinc to protect against corrosion.

GALVANIZED ROPE: Rope made of galvanized wire.

GALVANIZED STRAND: Strand made of galvanized wire.

GALVANIZED WIRE: Wire coated with zinc.

GANTRY (A-frame): A structural frame, extending above the superstructure, to which the boom support ropes are reeved.

GROMMET: A seven-strand wire-rope sling made from one continuous length of strand or an endless synthetic-web sling.

GROOVED DRUM: Drum with grooved outer surface to accommodate and guide a rope.

GROOVES: Depressions in the outer surface of a sheave or drum for positioning and supporting a rope.

GUY LINE: Strand or rope, usually galvanized, for holding a structure in a fixed position.

H

HANDLE: A terminal fitting to which metal mesh fabric is attached.

HANDLE EYE: An opening in a handle of a metal mesh sling shaped to accept a hook, shackle, or other lifting device.

HANDLING FIXTURE: A cradle, structure, shipping fixture, or container designed specifically to facilitate supporting, lifting, or handling a component during fabrication, loading, shipping, storage, or installation.

HITCH: A sling configuration whereby the sling is fastened to an object or load, either directly to it or around it.

HOIST (or hoisting): An apparatus that may be part of a crane, exerting a force for lifting or lowering or meaning all crane or derrick functions such as lowering, lifting, swinging, booming in and out or up and down, or suspending a personnel platform.

HOIST CHAIN: The load bearing chain in a hoist.

HOIST MOTION: That motion of a crane that raises and lowers a load.

HOOK LOAD: The total live weight supported by the hook of a crane, derrick, or other hoisting equipment, including the load, slings, spreader bars, and other tackle not part of the load but supported by the hook and required for the handling of the load.

I

IDLER: Sheave or roller used to guide or support a rope.

IMPROVED PLOW STEEL ROPE: A specific grade of wire rope.

INDEPENDENT WIRE-ROPE CORE (IWRC): Wire rope used as the axial member (core) of a larger wire rope.

INNER WIRES: All wires of a strand except the surface or cover wires.

INTERNALLY LUBRICATED: Wire rope or strand having all wires coated with lubricants.

J

JIB: An extension attached to the boom point to provide added boom length for lifting specified loads. The jib may be in line with the boom or offset to various angles.

K

KINK: Permanent distortion of wires and strands resulting from sharp bends or being pulled down too tight. It represents irreparable damage to and an indeterminate loss of strength in the rope.

L

LAGGING: External wood covering on a reel of rope or a strand or the grooved shell of a drum.

LAY LENGTH: The lengthwise distance on a wire rope in which a strand makes one complete turn around the rope's axis (see Figure 3.3).

LAY TYPES:

Left Lay: The direction of strand in which the cover wires are laid in a helix having a left-hand pitch, similar to a left-hand screw.

Right Lay: The direction of strand in which the cover wires are laid in a helix having a right-hand pitch, similar to a right-hand screw.

Regular Lay: The type of rope wherein the lay of the wire in the strand is in the opposite direction to the lay of the strand in the rope. The crowns of the wires appear to be parallel to the axis of the rope.

Lang Lay: The type of rope in which the lay of the wires in the strand is in the same direction as the lay of the strand in the rope. The crowns of the wires appear to be at an angle to the axis of the rope.

LIFT, CRITICAL: Lifting of parts, components, assemblies, or other items designated as critical because the effect of dropping, upset, or collision of them could:

- Present a potentially unacceptable risk of personnel injury or property damage.
- Result in significant release of radioactivity or other undesirable conditions.
- Cause undetectable damage resulting in future operational or safety problems.
- Cause significant work delay.

LIFT, ORDINARY: Any lift not designated as a critical lift or a pre-engineered production lift.

LIFT, PRE-ENGINEERED PRODUCTION: Repetitive, production-type lifting operation, independent of the nature of the load to be lifted, in which the probability of dropping, upset, or collision is reduced to a level acceptable to the responsible manager by preliminary engineering evaluation, specialized lifting fixtures, detailed procedures, operation-specific training, and independent review and approval of the entire process.

LINE: A rope used for supporting and controlling a suspended load.

LINK: A single ring of a chain.

LOAD: The total weight superimposed on the load block or hook.

LOAD BLOCK (lower): The assembly of hook or shackle, swivel, bearing, sheaves, pins, and frame suspended by the hoisting ropes.

LOAD BLOCK (upper): The assembly of hook or shackle, swivel, sheaves, pins, and frame suspended from the boom point.

LOAD REFUSAL: The point where the ultimate strength is exceeded.

LOAD HOIST: A hoist drum and rope reeving system used for hoisting and lowering loads.

LOAD RATINGS: Crane ratings in pounds established by the manufacturer in accordance with paragraph (c) of 1910.179.

LOAD-BEARING PARTS: Any part of a material-handling device in which the induced stress is influenced by the hook load. *A primary* load-bearing part is a part the failure of which could result in dropping, upset, or uncontrolled motion of the load. Load-bearing parts which, if failed, would result in no more than stoppage of the equipment without causing dropping, upset, or loss of control of the load are not considered to be primary load-bearing parts.

LOOP: A 360° change of direction in the course of a wire rope which when pulled down tight will result in a kink.

M

MAGNET: An electromagnetic device carried on a crane hook and used to pick up loads magnetically.

MAIN HOIST: The hoist mechanism provided for lifting the maximum-rated load.

MALE HANDLE (triangle): A handle with a handle eye.

MAN TROLLEY: A trolley having an operator's cab attached to it.

MARLINE SPIKE: Tapered steel pin used in splicing wire rope.

MASTER COUPLING LINK: An alloy steel welded coupling link used as an intermediate link to join alloy steel chain to master links.

MASTER LINK OR GATHERING RING: A forged or welded steel link used to support all members (legs) of an alloy steel chain sling or wire rope sling.

MAXIMUM INTENDED LOAD: The total load of all employees, tools, materials, and other loads reasonably anticipated to be applied to a personnel platform or personnel platform component at any one time.

MECHANICAL: A method of control by friction.

MECHANICAL COUPLING LINK: A non-welded, mechanically-closed steel link used to attach master links, hooks, etc., to alloy steel chain.

MODULUS OF ELASTICITY: Mathematical quantity expressing the ratio, within the elastic limit, between a definite range of unit stress on a wire rope and the corresponding elongation.

MOUSING: A method of bridging the throat opening of a hook to prevent the release of load lines and slings, under service or slack conditions, by wrapping with soft wire, rope, heavy tape, or similar materials.

N

NON-ROTATING WIRE ROPE: See Rotation-Resistant Wire Rope.

O

OPEN SOCKET: A wire-rope fitting consisting of a basket and two ears with a pin.

OUTRIGGERS: Extendable or fixed metal arms, attached to the mounting base, which rest on supports at the outer ends of the crane.

P

PEENING: Permanent distortion of outside wire in a rope caused by pounding.

PLOW STEEL ROPE: A specific grade of wire rope.

PREFORMED WIRE ROPE: Wire rope in which the strands are permanently shaped, before being fabricated into the rope, to the helical form they assume in the wire rope.

PREFORMED STRAND: Strand in which the wires are permanently shaped, before being fabricated into the strands, to the helical form they assume in the strand.

PRE-STRESSING: Stressing a wire rope or strand before use under such a tension and for such a time that stretch that would otherwise occur once the load is picked up is largely nonexistent.

PROOF LOAD: The load applied in performance of a proof test.

PROOF TEST: A nondestructive tension test performed by the manufacturer or an equivalent entity to verify construction and workmanship of slings or rigging accessories.

Q

QUALIFIED: A person who, by possession of a recognized degree, certificate, or professional standing, or who, by extensive knowledge, training, and experience, has successfully demonstrated an ability and competence to solve or resolve problems relating to the subject matter and work.

QUALIFIED ENGINEER/QUALIFIED ENGINEERING ORGANIZATION: An engineer or engineering organization whose competence in evaluation of the type of equipment in question has been demonstrated to the satisfaction of the responsible manager.

QUALIFIED INSPECTOR: One whose competence is recognized by the responsible manager and whose qualification to perform specific inspection activities has been determined, verified, and attested to in writing.

QUALIFIED OPERATOR: One who has had appropriate and approved training, including satisfactory completion of both written and operational tests to demonstrate knowledge, competence, and skill, in the safe operation of the equipment to be used.

QUALIFIED RIGGER: One whose competence in this skill has been demonstrated by experience satisfactory to the employer or appointed person.

R

RATED CAPACITY: The maximum hook load that a piece of hoisting equipment is designed to carry; also the maximum load that a sling, hook, shackle, or other rigging tackle is designed to carry.

> *Note: At the option of the user, a rated capacity can be assigned that is less than the design-rated capacity.*

RATED LOAD: The maximum load for which a crane or individual hoist is designed and built by the manufacturer and shown on the equipment nameplate(s).

RATED CAPACITY OR WORKING LOAD LIMIT: The maximum working load permitted by the provisions of 1910.184.

REACH: The effective length of an alloy steel chain sling measured from the top bearing surface of the upper terminal component to the bottom bearing surface of the lower terminal component.

REGENERATIVE: A form of dynamic braking in which the electrical energy generated is feedback into the power system.

REEL: The flanged spool on which wire rope or strand is wound for storage or shipment.

REEVING: A system in which a rope travels around drums or sheaves.

REGULAR-LAY ROPE: Wire rope in which the wires in the strands and the strands in the rope are laid in opposite directions.

REVERSE BEND: Reeving of a wire rope over sheaves and drums so that it bends in opposing directions.

RIGGING: The hardware or equipment used to safely attach a load to a lifting device. The art or process of safely attaching a load to a hook by means of adequately rated and properly applied slings and related hardware.

ROLLERS: Relatively small-diameter cylinders or wide-faced sheaves used for supporting or guiding ropes.

ROPE: Refers to wire rope, unless otherwise specified.

ROTATION-RESISTANT WIRE ROPE: Wire rope consisting of a left-lay, lang-lay inner rope covered by right-lay, regular-lay outer strands. This has the effect of counteracting torque by reducing the tendency of new rope to rotate.

RUNNING SHEAVE: A sheave that rotates as the load block is raised or lowered.

RUNWAY: Assembly of rails, girders, brackets, and framework on which a crane operates. A firm, level surface designed, prepared and designated as a path of travel for the weight and configuration of the crane being used to lift and travel with the crane suspended platform. An existing surface may be used as long as it meets these criteria.

S

SAFE WORKING LOAD: Load that a rope may carry economically and safely. This is a valid term only when the rope is new and equipment is in good condition.

SEALE: A strand construction having one size of cover wires with the same number of one size of wires in the inner layer and each layer having the same length and direction of lay. Most common construction is one center wire, nine inner wires, and nine cover wires.

SEIZE: To securely bind the end of a wire rope or strand with seizing wire or strand.

SEIZING STRAND: Small strand, usually of seven wires, made of soft-annealed-iron wire.

SEIZING WIRE: A soft-annealed iron wire.

SELVAGE EDGE: The finished edge of synthetic webbing designed to prevent unraveling.

SERVE: To cover the surface of a wire rope or strand with a wrapping of wire.

SHACKLE: A type of clevis normally used for lifting.

SHALL: A word indicating that an action is mandatory.

SHEAVE: A grooved wheel or pulley used with a rope to change direction and point of application of a pulling force.

SHEAVE, NON-RUNNING (equalizer): A sheave used to equalize tension in opposite parts of a rope, called non-running because of its slight movement.

SHEAVE, RUNNING: A sheave that rotates as the load block is lifted or lowered.

SHOCK LOAD: A force that results from the rapid application of a force (such as impacting or jerking) or rapid movement of a static load. A shock load significantly adds to the static load.

SHOULD: A word indicating a recommended action, the advisability of which depends on the facts in each situation.

SIDE LOADING: A load applied at an angle to the vertical plane of the boom.

SIDE PULL: That portion of a hoist pull acting horizontally when the hoist lines are not operated vertically.

SLING MANUFACTURER: A person or organization that assembles sling components into their final form for sale to users.

SLING/SLINGS: An assembly of wire ropes, chains, synthetic web, and metal mesh made into forms, with or without fittings and used to connect the load to the material handling equipment.

SLINGS, BRAIDED: Very flexible slings composed of two or more individual wire ropes braided together.

SMOOTH-FACED DRUM: Drum with a plain, not grooved, face.

SOCKET: Generic name for a type of wire rope fitting.

SPAN: The horizontal, center-to-center distance of runway rails.

SPIRAL: A single transverse coil that is the basic element from which metal mesh is fabricated.

SPIRAL GROOVE: Groove that follows the path of a helix around a drum, similar to the thread of a screw thread.

SPLICING: Interweaving of two ends of rope to make a continuous or endless length without appreciably increasing the diameter. Also refers to making a loop or eye in the end of a rope by tucking the ends of the strands back into the main body of the rope.

Splice, Hand Tucked: A loop or eye formed in the end of a rope by tucking the end of the strands back into the main body of the rope in a prescribed manner.

Splice, Mechanical: A loop or eye formed in the end of a wire rope by pressing or swaging one or more metal sleeve over the wire rope junction.

STAINLESS-STEEL ROPE: Wire rope made of chrome-nickel steel wires having great resistance to corrosion.

STANDBY CRANE: A crane that is not in regular service but that is used occasionally or intermittently as required.

STANDING ROPE (guy): A supporting rope which maintains a constant distance between the points of attachment to the two components connected by the rope.

STATIC LOAD: The load resulting from a constant applied force or load.

STEEL-CLAD ROPE: Rope with individual strands spirally wrapped with flat steel wire.

STOP: A device to limit travel of a trolley or crane bridge. This device normally is attached to a fixed structure and normally does not have energy absorbing ability.

STRAND: An arrangement of wires helically laid about an axis or another wire or fiber center to produce a symmetrical section.

STRAND LAID ENDLESS SLING-MECHANICAL JOINT: A wire rope sling made endless from one length of rope with the ends joined by one or more metallic fittings.

STRAND LAID GROMMET-HAND TUCKED: An endless wire rope sling made from one length of strand wrapped six times around a core formed by hand tucking the ends of the strand inside the six wraps.

STRAND LAID ENDLESS SLING-MECHANICAL JOINT: A wire rope sling made endless from one length of rope with the ends joined by one or more metallic fittings.

STRAND LAID ROPE: A wire rope made with strands (usually six or eight) wrapped around a fiber core, wire strand core, or independent wire rope core (IWRC).

STRESS: The force or resistance within any solid body against alteration of form; in the case of a solid wire it would be the load on the rope divided by the cross-section of the wire.

STRETCH: The elongation of a wire rope under load.

STRUCTURAL COMPETENCE: The ability of the machine and its components to withstand the stresses imposed by applied loads.

SUPERSTRUCTURE: The rotating upper frame structure of the machine and the operating machinery mounted thereon.

SWAGED FITTINGS: Fittings in which wire rope can be inserted and then permanently attached by a cold-forming method (swaging) the shank that encloses the rope.

SWING: The rotation of the superstructure for movement of loads in a horizontal direction about the axis of rotation.

SWING MECHANISM: The machinery involved in providing rotation of the superstructure.

SWITCH, ELECTRIC: A device for making, breaking, or changing the connections in an electrical circuit.

SWITCH, EMERGENCY STOP: A manually or automatically operated electric switch to cut off electric power independently of the regular operating controls.

SWITCH, LIMIT: A switch that is operated by some part or motion of a power-driven machine or equipment to alter the electrical circuit associated with the machine or equipment.

SWITCH, MAIN: A switch controlling the entire power supply to a crane or other equipment, often called the disconnect switch.

SWITCH, MASTER: A switch which dominates the operation of contactors, relays, or other remotely operated devices.

T

TACKLE: An assembly of ropes and sheaves arranged for hoisting and pulling.

TAG LINE: A rope used to prevent rotation and swinging of a load.

TAPERING AND WELDING: Reducing the diameter of the end of a wire rope and welding it to facilitate reeving.

THIMBLE: Grooved metal fitting to protect the eye of a wire rope.

TIERING: The process of placing one load on or above another.

TINNED WIRE: Wire coated with tin.

TRANSIT: The moving or transporting of a crane from one jobsite to another.

TRAVEL: The function of the machine moving from one location to another, on a jobsite.

TRAVEL MECHANISM: The machinery involved in providing travel.

TROLLEY: A unit consisting of frame, trucks, trolley drive, and hoisting mechanism moving on the bridge rails in a direction at right angles to the crane runway.

TROLLEY GIRTS: Structural members that are supported on the trolley trucks and that contain the upper sheave assemblies.

TROLLEY TRAVEL: Horizontal travel of a trolley at right angles to runway rails.

TRUCK: The unit consisting of a frame, wheels, bearings, and axles and which supports the bridge girders or trolleys.

TWO-BLOCKING: The act of continued hoisting in which the load-block and head-block assemblies are brought into physical contact, thereby preventing further movement of the load block and creating shock loads to the rope and reeving system.

U

ULTIMATE LOAD: The average load or force at which the equipment/product fails or no longer supports the load.

V

VERIFICATION: A procedure in which a design, calculation, drawing, procedure, instruction, report, or document is checked and signed by one or more parties. The one or more persons designated to sign verify, based on personal observation, certified records, or direct reports, that a specific action has been performed in accordance with specified requirements.

VERTICAL HITCH: A method of supporting a load by a single, vertical part, or leg of the sling.

W

WARRINGTON: The name for a type of strand pattern that is characterized by having one of its wire layers (usually the outer) made up of an arrangement of alternately large and small wires.

WEDGE SOCKET: Wire rope fittings wherein the rope end is secured by a wedge.

WHEEL BASE: Distance between centers of outermost wheels for bridge and trolley trucks. The distance between centers of front and rear axles. For a multiple axle assembly, the axle center for wheelbase measurement is taken as the midpoint of the assembly.

WHEEL LOAD: The load on any wheel with the trolley and lifted load (rated load) positioned on the bridge to give maximum-loading conditions.

WHIPLINE (auxiliary hoist): A separate hoist rope system of lighter load capacity and higher speed than provided by the main hoist.

WINCH HEAD: A power driven spool for handling of loads by means of friction between fiber or wire rope and spool.

WIRE ROPE: A plurality of wire strands laid helically around an axis or a core.

WIRE (round): Single continuous length of metal, cold drawn from a rod.

WIRE STRAND CORE (WSC): A wire strand used as the axial member of a wire rope.

WORKING LOAD: The external load, in pounds, applied to the crane, including the weight of load-attaching equipment such as load blocks, shackles, and slings that the product is authorized to support in a particular service.

WORKING LOAD LIMIT: The maximum mass of force that the equipment/product is authorized to support in general service when the pull is applied in-line, unless noted otherwise, with respect to the centerline of the product. The term is used interchangeably with the following terms:

- ▸ WLL
- ▸ Rated Load Value
- ▸ Resultant Working Load.

APPENDIX A

OSHA STANDARD 1910.6
INCORPORATION BY REFERENCE

The following Standard 29 CFR 1910.6 provides an explanation of how standards of other U.S. Government agencies and organizations which are not agencies of the U.S. Government are adopted as standards under the Occupational Safety and Health Act and have the same force and effect as other standards in this part.

Incorporation by Reference

Standard Number: 1910.6

Standard Title: Incorporation by reference.

SubPart Number: A

SubPart Title: General

(a)(1)

The standards of agencies of the U.S. Government, and organizations which are not agencies of the U.S. Government which are incorporated by reference in this part, have the same force and effect as other standards in this part. Only the mandatory provisions (i.e., provisions containing the word "shall" or other mandatory language) of standards incorporated by reference are adopted as standards under the Occupational Safety and Health Act.

(a)(2)

Any changes in the standards incorporated by reference in this part and an official historic file of such changes are available for inspection at the national office of the Occupational Safety and Health Administration, U.S. Department of Labor, Washington, DC 20210.

(a)(3)

The materials listed in paragraphs (b) through (w) of this section are incorporated by reference in the corresponding sections noted as they exist on the date of the approval, and a notice of any change in these materials will be published in the Federal Register. These incorporations by reference were approved by the Director of the Federal Register in accordance with 5 U.S.C. 552(a) and 1 CFR part 51.

(a)(4)

Copies of the following standards that are issued by the respective private standards organizations may be obtained from the issuing organizations. The materials are available for purchase at the corresponding addresses of the private standards organizations noted below. In addition, all are available for inspection at the Office of the Federal Register, 800 North Capitol Street, NW., Suite 700, Washington DC, and through the OSHA Docket Office, Room N2625, U.S. Department of Labor, 200 Constitution Ave., Washington, DC 20210, or any of its regional offices.

(e)

The following material is available for purchase from the American National Standards Institute (ANSI), 11 West 42nd St., New York, NY 10036:

(e)(18)

ANSI B30.2-43 7 52) Safety Code for Cranes, Derricks, and Hoists, IBR approved for Sec. 1910.261(a)(3)(xi), (c)(2)(vi), and (c)(8)(i) and (iv).

(e)(19)

ANSI B30.2.0-67 Safety Code for Overhead and Gantry Cranes, IBR approved for Secs. 1910.179(b)(2); 1910.261(a)(3)(xii), (c)(2)(v), and (c)(8)(i) and (iv).

(e)(20)

ANSI B30.5-68 Safety Code for Crawler, Locomotive, and Truck Cranes, IBR approved for Secs. 1910.180(b)(2) and 1910.261(a)(3)(xiii).

APPENDIX B

ADVANCE NOTICE OF PROPOSED RULEMAKING IN CRANE SAFETY

Information Date: 10/19/1992

Federal Register #: 57:47746

Standard Number: 1910

Type: Proposed

Agency: OSHA

Subject: Advance Notice of Proposed Rulemaking in Crane Safety

CFR Title: 29

Abstract: OSHA is considering a multi-phased revision of the crane safety provisions of 29 CFR Part 1926, Subpart N (Cranes, Derricks, Hoists, Elevators and Conveyors) and of 29 CFR 1910, Subpart N, (Materials Handling and Storage). One of the primary areas of concern to the Agency is the limited criteria for crane operator qualifications incorporated by reference into the existing regulations. Other areas OSHA would explore and evaluate include: The need to update parts of the standard dealing with the use, inspection, and maintenance of cranes; the need for a requirement for certification of cranes used on construction sites and general industry sites; and the need for a requirement for certification of riggers and signal persons. OSHA is soliciting quantitative and qualitative data, expert opinions, comments and information regarding crane safety in general, and crane operation qualifications in particular. This information will allow the Agency to evaluate the need for stricter crane operator qualifications criteria, and aid in the development of any other revisions to the existing crane standards may be appropriate. Written comments on the advance notice of proposed rulemaking must be postmarked by February 12, 1993.

Comments and information should be submitted in quadruplicate to the Docket Officer, Docket No. S-400 OSHA, room N2634, USDOL, 200 Constitution Ave, NW, Washington, DC 20210. Telephone (202) 523-7894. For further information contact Mr. James Foster, OSHA, USDOL, room N3637, 200 Constitution Avenue, NW., Washington, DC 20210, Telephone (202) 523-8151.

2173. CRANE SAFETY

Legal Authority: 29 USC 655(b); 40 USC 333; 33 USC 941

CFR Citation: 29 CFR 1926.550; 29 CFR 1926.552; 29 CFR 1926.553; 29 CFR 1926.554; 29 CFR 1926.556; 29 CFR 1910.67; 29 CFR 1910.179; 29 CFR 1910.180; 29 CFR 1919.181

Legal Deadline: None

Abstract: The present crane regulations for construction and general industry have not been revised since being promulgated in 1971. They rely heavily on outdated 1968 ANSI standards. OSHA has received comments that the existing provisions are inadequate and need revision to reflect current conditions and equipment. It has also been suggested that there is a need to establish additional crane installation and use provisions, including possible certification programs for crane operators and riggers.

Timetable:

Action	Date	FR Cite
ANPRM	10/19/92	57 FR 47746

ANPRM Comment Period End:	02/12/93
Withdrawn:	03/31/95
Small Entities Affected:	None
Government Levels Affected:	None

Agency Contact: Thomas H. Seymour, Acting Director, Safety Standards Programs, Department of Labor, Occupational Safety and Health Administration, 200 Constitution Avenue NW., Room N3605, FP Building, Washington, DC 20210, 202 219-8061

RIN: 1218-AB38

APPENDIX C

STATES WITH APPROVED PLANS

The ultimate accreditation of a state's plan is called "final approval." When OSHA grants final approval to a state under section 18(e) of the Act, it relinquishes its authority to cover occupational safety and health matters covered by the state. After at least one year following certification, the state becomes eligible for final approval if OSHA determines that it is providing, in actual operation, worker protection "at least as effective" as the protection provided by the federal program. The state also must meet 100 percent of the established compliance staffing levels (benchmarks) and participate in OSHA's computerized inspection data system before OSHA can grant final approval.

Section 18 of the Occupational Safety and Health Act of 1970 (the Act) encourages states to develop and operate their own job safety and health programs. OSHA approves and monitors state plans and provides up to 50 percent of an approved plan's operating costs.

States must set job safety and health standards that are "at least as effective as" comparable federal standards. (Most states adopt standards identical to federal ones). States have the option to promulgate standards covering hazards not addressed by federal standards.

A state must conduct inspections to enforce its standards, cover public (state and local government) employees, and operate occupational safety and health training and education programs. In addition, most states provide free on-site consultation to help employers identify and correct workplace hazards. Such consultation may be provided either under the plan or through a special agreement under section 7 (c)(1) of the Act.

To gain OSHA approval for a "developmental plan," the first step in the state plan process, a state must assure OSHA that within three years it will have in place all the structural elements necessary for an effective occupational safety and health program. These elements include appropriate legislation; regulations and procedures for standards setting, enforcement, and appeal of citations and penalties; and a sufficient number of qualified enforcement personnel.

There are currently 23 states and jurisdictions operating complete state plans (covering both theprivate sector and state and local government employees) and two, Connecticut and New York,which cover public employees only. Eight other states were approved at one time but subsequently withdrew their programs.

States with occupational safety and health programs:

Alaska	Michigan	Tennessee
Arizona	Minnesota	Utah
California	Nevada	Vermont
Connecticut	New Mexico	Virgin Islands
Hawaii	New York	Virginia
Indiana	North Carolina	Washington
Iowa	Oregon	Wyoming
Kentucky	Puerto Rico	
Maryland	South Carolina	

(Connecticut and New York plans cover public sector only.)

Once a state has completed and documented all its developmental steps, it is eligible for certification. Certification renders no judgment as to actual state performance, but merely attests to the structural completeness of the plan. Twenty-four states have received certification.

At any time after initial plan approval, when it appears that the state is capable of independently enforcing standards, OSHA may enter into an "operational status agreement" with the state. This commits OSHA to suspend the exercise of discretionary federal enforcement in all or certain activities covered by the state plan.

The ultimate accreditation of a state's plan is called "final approval." When OSHA grants final approval to a state under section 18 (e) of the Act, it relinquishes its authority to cover occupational safety and health matters covered by the state. After at least one year following certification, the state becomes eligible for final approval if OSHA determines that it is providing, in actual operation, worker protection "at least as effective" as the protection provided by the federal program. The state also must meet 100 percent of the established compliance staffing levels (benchmarks) and participate in OSHA's computerized inspection data system before OSHA can grant final approval.

Employees finding workplace safety and health hazards may file a formal complaint with the appropriate plan state or with the appropriate OSHA regional administrator. Complaints will be investigated and should include the name of the workplace, type(s) of hazard(s) observed, and any other pertinent information.

Anyone finding inadequacies or other problems in the administration of a state's program may file a Complaint About State Program Administration (CASPA) with the appropriate OSHA regional administrator as well. The complainant's name is kept confidential. OSHA investigates all such complaints, and where complaints are found to be valid, requires appropriate corrective action on the part of the state.

APPENDIX D

OSHA HAZARD INFORMATION BULLETIN - TRUCK CRANES

Information Date: 1989, 05-02

Record Type: Hazard Information Bulletin

Subject: Truck Cranes

May 2, 1989

MEMORANDUM FOR: REGIONAL ADMINISTRATORS

THRU: LEO CAREY

               Director

               Office of Field Programs

FROM: EDWARD BAIER

               Director

               Directorate of Technical Support

SUBJECT: Safety Hazard Information Bulletin on Truck Cranes

The Concord, New Hampshire, Area Office has brought to our attention a potentially serious hazard existing with the use of truck cranes with possibly insufficient load capacities. The problem occurs when trucks fitted with boom cranes are not specifically designed for such applications. The only available load capacity rating in such a situation is the rating of the crane boom structure itself. This rating is inappropriate for use with a truck-crane system since it does not take into account the size of the truck, strength of the truck platform, size or presence of outriggers, and tipping moment and other engineering mechanics considerations that would be required to evaluate and rate the total truck-crane package.

OSHA 1926.550(a)(1) requires that when manufacturer's specifications applicable to a crane are not available, limitations assigned to the equipment shall be based on the determinations of a qualified engineer. These specifications must take into consideration both the vehicle and crane characteristics. The American National Standards Institute (ANSI) A92.2-1979 National Standard for Vehicle-mounted Elevating and Rotating Aerial Devices applies to the design, construction, testing, inspection, care and use of machinery including truck cranes with extensible or articulating booms. The standard sets necessary load capacity specifications for both the crane and truck components of such units along with standards for the design and manufacture, testing and inspection, and training of operators for truck cranes.

Furthermore, the American Society of Mechanical Engineers (ASME)/ANSI B3O.22-1987 Standard on Articulating Boom Cranes sets load ratings for truck-mounted articulating boom cranes of one ton or greater capacity. ASME/ANSI B30.5-1982 Standard on Mobile and boom cranes. Load ratings where stability governs lifting performance are determined by adherence to the ANSI Society of Automotive Engineers (SAE 765a) October, 1980 Crane Load Stability Test Code.

APPENDIX E

OSHA STANDARD 1910.27 - FIXED LADDERS

Standard Number: 1910.27

Standard Title: Fixed ladders.

SubPart Number: D

SubPart Title: Walking-Working Surfaces

(a)

"Design requirements" -

(a)(1)

Design considerations. All ladders, appurtenances, and fastenings shall be designed to meet the following load requirements:

(a)(1)(i)

The minimum design live load shall be a single concentrated load of 200 pounds.

(a)(1)(ii)

The number and position of additional concentrated live-load units of 200 pounds each as determined from anticipated usage of the ladder shall be considered in the design.

(a)(1)(iii)

The live loads imposed by persons occupying the ladder shall be considered to be concentrated at such points as will cause the maximum stress in the structural member being considered.

(a)(1)(iv)

The weight of the ladder and attached appurtenances together with the live load shall be considered in the design of rails and fastenings.

(a)(2)

"Design stresses." Design stresses for wood components of ladders shall not exceed those specified in 1910.25. All wood parts of fixed ladders shall meet the requirements of 1910.25(b).

For fixed ladders consisting of wood side rails and wood rungs or cleats, used at a pitch in the range 75 degrees to 90 degrees, and intended for use by no more than one person per section, single ladders as described in 1910.25(c)(3)(ii) are acceptable.

(b)

"Specific features" -

(b)(1)

"Rungs and cleats."

(b)(1)(i)

All rungs shall have a minimum diameter of three-fourths inch for metal ladders, except as covered in paragraph (b)(7)(i) of this section and a minimum diameter of 1 1/8 inches for wood ladders.

(b)(1)(ii)

The distance between rungs, cleats, and steps shall not exceed 12 inches and shall be uniform throughout the length of the ladder.

(b)(1)(iii)

The minimum clear length of rungs or cleats shall be 16 inches.

(b)(1)(iv)

Rungs, cleats, and steps shall be free of splinters, sharp edges, burrs, or projections which may be a hazard.

(b)(1)(v)

The rungs of an individual-rung ladder shall be so designed that the foot cannot slide off the end. A suggested design is shown in figure D-1.

(b)(2)

"Side rails." Side rails which might be used as a climbing aid shall be of such cross sections as to afford adequate gripping surface without sharp edges, splinters, or burrs.

(b)(3)

"Fastenings." Fastenings shall be an integral part of fixed ladder design.

(b)(4)

"Splices." All splices made by whatever means shall meet design requirements as noted in paragraph (a) of this section. All splices and connections shall have smooth transition with original members and with no sharp or extensive projections.

(b)(5)

"Electrolytic action." Adequate means shall be employed to protect dissimilar metals from electrolytic action when such metals are joined.

(b)(6)

"Welding." All welding shall be in accordance with the "Code for Welding in Building Construction" (AWSD1.0-1966).

(b)(7)

"Protection from deterioration."

(b)(7)(i)

Metal ladders and appurtenances shall be painted or otherwise treated to resist corrosion and rusting when location demands. Ladders formed by individual metal rungs imbedded in concrete, which serve as access to pits and to other areas under floors, are frequently located in an atmosphere that causes corrosion and rusting. To increase rung life in such atmosphere, individual metal rungs shall have a minimum diameter of 1 inch or shall be painted or otherwise treated to resist corrosion and rusting.

(b)(7)(ii)

Wood ladders, when used under conditions where decay may occur, shall be treated with a nonirritating preservative, and the details shall be such as to prevent or minimize the accumulation of water on wood parts.

(b)(7)(iii)

When different types of materials are used in the construction of a ladder, the materials used shall be so treated as to have no deleterious effect one upon the other.

(c)

"Clearance" -

(c)(1)

"Climbing side." On fixed ladders, the perpendicular distance from the centerline of the rungs to the nearest permanent object on the climbing side of the ladder shall be 36 inches for a pitch of 76 degrees, and 30 inches for a pitch of 90 degrees (fig. D-2 of this section), with minimum clearances for intermediate pitches varying between these two limits in proportion to the slope, except as provided in subparagraphs (3) and (5) of this paragraph.

(c)(2)

"Ladders without cages or wells." A clear width of at least 15 inches shall be provided each way from the centerline of the ladder in the climbing space, except when cages or wells are necessary.

(c)(3)

"Ladders with cages or baskets." Ladders equipped with cage or basket are excepted from the provisions of subparagraphs (1) and (2) of this paragraph, but shall conform to the provisions of paragraph (d)(1)(v) of this section. Fixed ladders in smooth-walled wells are excepted from the provisions of subparagraph (1) of this paragraph, but shall conform to the provisions of paragraph (d)(1)(vi) of this section.

(c)(4)

"Clearance in back of ladder." The distance from the centerline of rungs, cleats, or steps to the nearest permanent object in back of the ladder shall be not less than 7 inches, except that when unavoidable obstructions are encountered, minimum clearances as shown in figure D-3 shall be provided.

(c)(5)

"Clearance in back of grab bar." The distance from the centerline of the grab bar to the nearest permanent object in back of the grab bars shall be not less than 4 inches. Grab bars shall not protrude on the climbing side beyond the rungs of the ladder which they serve.

(c)(6)

"Step-across distance." The step-across distance from the nearest edge of ladder to the nearest edge of equipment or structure shall be not more than 12 inches, or less than 2 2 inches (fig. D-4).

(c)(7)

"Hatch cover." Counterweighted hatch covers shall open a minimum of 60 degrees from the horizontal. The distance from the centerline of rungs or cleats to the edge of the hatch opening on the climbing side shall be not less than 24 inches for offset wells or 30 inches for straight wells. There shall be not protruding potential hazards within 24 inches of the centerline of rungs or cleats; any such hazards within 30 inches of the centerline of the rungs or cleats shall be fitted with deflector plates placed at an angle of 60 degrees from the horizontal as indicated in figure D-5. The relationship of a fixed ladder to an acceptable counterweighted hatch cover is illustrated in figure D-6.

(d)

"Special requirements" -

(d)(1)

"Cages or wells."

(d)(1)(i)

Cages or wells (except on chimney ladders) shall be built, as shown on the applicable drawings, covered in detail in figures D-7, D-8, and D-9, or of equivalent construction.

(d)(1)(ii)

Cages or wells (except as provided in subparagraph (5) of this paragraph) conforming to the dimensions shown in figures D-7, D-8, and D-9 shall be provided on ladders of more than 20 feet to a maximum unbroken length of 30 feet.

(d)(1)(iii)

Cages shall extend a minimum of 42 inches above the top of landing, unless other acceptable protection is provided.

(d)(1)(iv)

Cages shall extend down the ladder to a point not less than 7 feet nor more than 8 feet above the base of the ladder, with bottom flared not less than 4 inches, or portion of cage opposite ladder shall be carried to the base.

(d)(1)(v)

Cages shall not extend less than 27 nor more than 28 inches from the centerline of the rungs of the ladder. Cage shall not be less than 27 inches in width. The inside shall be clear of projections. Vertical bars shall be located at a maximum spacing of 40 degrees around the circumference of the cage; this will give a maximum spacing of approximately 9 1/2 inches, center to center.

(d)(1)(vi)

Ladder wells shall have a clear width of at least 15 inches measured each way from the centerline of the ladder. Smooth-walled wells shall be a minimum of 27 inches from the centerline of rungs to the well wall on the climbing side of the ladder. Where other obstructions on the climbing side of the ladder exist, there shall be a minimum of 30 inches from the centerline of the rungs.

(d)(2)

"Landing platforms." When ladders are used to ascend to heights exceeding 20 feet (except on chimneys), landing platforms shall be provided for each 30 feet of height or fraction thereof, except that, where no cage, well, or ladder safety device is provided, landing platforms shall be provided for each 20 feet of height or fraction thereof. Each ladder section shall be offset from adjacent sections. Where installation conditions (even for a short, unbroken length) require that adjacent sections be offset, landing platforms shall be provided at each offset.

(d)(2)(i)

Where a man has to step a distance greater than 12 inches from the centerline of the rung of a ladder to the nearest edge of structure or equipment, a landing platform shall be provided. The minimum step-across distance shall be 2 2 inches.

(d)(2)(ii)

All landing platforms shall be equipped with standard railings and toeboards, so arranged as to give safe access to the ladder. Platforms shall be not less than 24 inches in width and 30 inches in length.

(d)(2)(iii)

One rung of any section of ladder shall be located at the level of the landing laterally served by the ladder. Where access to the landing is through the ladder, the same rung spacing as used on the ladder shall be used from the landing platform to the first rung below the landing.

(d)(3)

"Ladder extensions." The side rails of through or side-step ladder extensions shall extend 3 2 feet above parapets and landings. For through ladder extensions, the rungs shall be omitted from the extension and shall have not less than 18 nor more than 24 inches clearance between rails. For side-step or offset fixed ladder sections, at landings, the side rails and rungs shall be carried to the next regular rung beyond or above the 3 2 feet minimum (fig. D-10).

(d)(4)

"Grab bars." Grab bars shall be spaced by a continuation of the rung spacing when they are located in the horizontal position. Vertical grab bars shall have the same spacing as the ladder side rails. Grab-bar diameters shall be the equivalent of the round-rung diameters.

(d)(5)

"Ladder safety devices." Ladder safety devices may be used on tower, water tank, and chimney ladders over 20 feet in unbroken length in lieu of cage protection. No landing platform is required in these cases. All ladder safety devices such as those that incorporate lifebelts, friction brakes, and sliding attachments shall meet the design requirements of the ladders which they serve.

(e)

"Pitch" -

(e)(1)

"Preferred pitch." The preferred pitch of fixed ladders shall be considered to come in the range of 75 degrees and 90 degrees with the horizontal (fig. D-11).

(e)(2)

"Substandard pitch." Fixed ladders shall be considered as substandard if they are installed within the substandard pitch range of 60 and 75 degrees with the horizontal. Substandard fixed ladders are permitted only where it is found necessary to meet conditions of installation. This substandard pitch range shall be considered as a critical range to be avoided, if possible.

(e)(3)

"Scope of coverage in this section." This section covers only fixed ladders within the pitch range of 60 degrees and 90 degrees with the horizontal.

(e)(4)

"Pitch greater than 90 degrees." Ladders having a pitch in excess of 90 degrees with the horizontal are prohibited.

(f)

"Maintenance." All ladders shall be maintained in a safe condition. All ladders shall be inspected regularly, with the intervals between inspections being determined by use and exposure.

APPENDIX F

OSHA STANDARD 1910.333(c)(3) - OVERHEAD LINES

1910.333(c)(3)

"Overhead lines."

If work is to be performed near overhead lines, the lines shall be deenergized and grounded, or other protective measures shall be provided before work is started. If the lines are to be deenergized, arrangements shall be made with the person or organization that operates or controls the electric circuits involved to deenergize and ground them. If protective measures, such as guarding, isolating, or insulating, are provided, these precautions shall prevent employees from contacting such lines directly with any part of their body or indirectly through conductive materials, tools, or equipment.

Note: The work practices used by qualified persons installing insulating devices on overhead power transmission or distribution lines are covered by 1910.269 of this Part, not by 1910.332 through 1910.335 of this Part. Under paragraph (c)(2) of this section, unqualified persons are prohibited from performing this type of work.

(c)(3)(i) "Unqualified persons."

(c)(3)(i)(A)

When an unqualified person is working in an elevated position near overhead lines, the location shall be such that the person and the longest conductive object he or she may contact cannot come closer to any unguarded, energized overhead line than the following distances:

(c)(3)(i)(A)(1)

For voltages to ground 50kV or below - 10 feet (305 cm);

(c)(3)(i)(A)(2)

For voltages to ground over 50kV - 10 feet (305 cm) plus 4 inches (10 cm) for every 10kV over 50kV.

(c)(3)(i)(B)

When an unqualified person is working on the ground in the vicinity of overhead lines, the person may not bring any conductive object closer to unguarded, energized overhead lines than the distances given in paragraph (c)(3)(i)(A) of this section.

Note: For voltages normally encountered with overhead power line, objects which do not have an insulating rating for the voltage involved are considered to be conductive.

(c)(3)(ii)

"Qualified persons." When a qualified person is working in the vicinity of overhead lines, whether in an elevated position or on the ground, the person may not approach or take any conductive object without an approved insulating handle closer to exposed energized parts than shown in Table S-5 unless:

(c)(3)(ii)(A)

The person is insulated from the energized part (gloves, with sleeves if necessary, rated for the voltage involved are considered to be insulation of the person from the energized part on which work is performed), or

(c)(3)(ii)(B)

The energized part is insulated both from all other conductive objects at a different potential and from the person, or

(c)(3)(ii)(C)

The person is insulated from all conductive objects at a potential different from that of the energized part.

(c)(3)(iii)

A Vehicular and mechanical equipment."

TABLE S-5. APPROACH DISTANCES FOR QUALIFIED EMPLOYEES — ALTERNATING CURRENT

Voltage range (phase to phase)	Minimum approach distance
300V and less	Avoid contact
Over 300V, not over 750V	1 ft. 0 in. (30.5 cm)
Over 750V, not over 2kV	1 ft. 6 in. (46 cm)
Over 2kV, not over 15kV	2 ft. 0 in. (61 cm)
Over 15kV, not over 37kV	3 ft. 0 in. (91 cm)
Over 37kV, not over 87.5kV	3 ft. 6 in. (107 cm)
Over 87.5kV, not over 121kV	4 ft. 0 in. (122 cm)
Over 121kV, not over 140kV	4 ft. 6 in. (137 cm)

(c)(3)(iii)(A)

Any vehicle or mechanical equipment capable of having parts of its structure elevated near energized overhead lines shall be operated so that a clearance of 10 ft. (305 cm) is maintained. If the voltage is higher than 50kV, the clearance shall be increased 4 in. (10 cm) for every 10kV over that voltage. However, under any of the following conditions, the clearance may be reduced:

(c)(3)(iii)(A)(1)

If the vehicle is in transit with its structure lowered, the clearance may be reduced to 4 ft. (122 cm). If the voltage is higher than 50kV, the clearance shall be increased 4 in. (10 cm) for every 10 kV over that voltage.

(c)(3)(iii)(A)(2)

If insulating barriers are installed to prevent contact with the lines, and if the barriers are rated for the voltage of the line being guarded and are not a part of or an attachment to the vehicle or its raised structure, the clearance may be reduced to a distance within the designed working dimensions of the insulating barrier.

(c)(3)(iii)(A)(3)

If the equipment is an aerial lift insulated for the voltage involved, and if the work is performed by a qualified person, the clearance (between the uninsulated portion of the aerial lift and the power line) may be reduced to the distance given in Table S-5.

(c)(3)(iii)(B)

Employees standing on the ground may not contact the vehicle or mechanical equipment or any of its attachments, unless:

(c)(3)(iii)(B)(1)

The employee is using protective equipment rated for the voltage; or

(c)(3)(iii)(B)(2)

The equipment is located so that no uninsulated part of its structure (that portion of the structure that provides a conductive path to employees on the ground) can come closer to the line than permitted in paragraph (c)(3)(iii) of this section.

(c)(3)(iii)(C)

If any vehicle or mechanical equipment capable of having parts of its structure elevated near energized overhead lines is intentionally grounded, employees working on the ground near the point of grounding may not stand at the grounding location whenever there is a possibility of overhead line contact. Additional precautions, such as the use of barricades or insulation, shall be taken to protect employees from hazardous ground potentials, depending on earth resistivity and fault currents, which can develop within the first few feet or more outward from the grounding point.

[55 FR 32016, Aug. 6, 1990; 55 FR 42053, Nov. 1, 1990; as amended at 59 FR 4476, Jan. 31, 1994]

APPENDIX G

NIOSH ALERT: OPERATING CRANES NEAR ELECTRICAL LINES

Occupational fatalities associated with electrocutions are a significant, ongoing problem. Data from the NIOSH National Traumatic Occupational Fatality (NTOF) surveillance system indicated that an average of 6,359 traumatic work-related deaths occurred each year in the United States from 1980 through 1989; an estimated 7% of these fatalities were due to electrocutions.[12] In 1995, the Bureau of Labor Statistics reported that electrocutions accounted for 6% of all worker deaths.[13] For the year 1990, the National Safety Council reported that electrocutions were the fourth leading cause of work-related traumatic death.[14]

Almost all American workers are exposed to electrical energy at sometime during their work day, and the same electrical hazards can affect workers in different industries. Based on an analysis of 224 cases, NIOSH identified five case scenarios that describe the incidents resulting in the fatalities: (1) direct worker contact with an energized power line (28%); (2) direct worker contact with energized equipment (21%); (3) boomed vehicle contact with an energized power line (18%); (4) improperly installed or damaged equipment (17%); (5) conductive equipment contact with an energized power line (16%). [17] From 1980 through 1989, NIOSH reported that an average of 15 electrocutions each year were caused by contact between cranes or some other type of boomed vehicles and energized, overhead power lines.

Each year, electrocutions represent 7% of injury-related fatalities.[1] NIOSH onsite investigators found that 13% of work-related electrocutions involved crane contact with overhead power lines. After evaluating the circumstances of these electrocutions, NIOSH disseminated two alerts that describe procedures and precautions around power lines. The following NIOSH Alert (Publication No. 95-108) was published in May 1995 and updates a previous NIOSH Alert published in July 1985 [NIOSH 1985].

Preventing Electrocutions of Crane Operators and Crew Members Working Near Overhead Power Lines

NIOSH ALERT: May 1995

DHHS (NIOSH) Publication No. 95-108

WARNING!

Crane operators and crew members may be electrocuted when they work near overhead power lines.

The National Institute for Occupational Safety and Health (NIOSH) requests assistance in preventing electrocutions of crane operators and crew members working near overhead power lines. Recent NIOSH investigations suggest that employers, supervisors, and workers may not be fully aware of the hazards of operating cranes near overhead power lines or may not implement the proper safety procedures for controlling these hazards. This alert describes five cases (six electrocutions) that resulted from such hazards and makes recommendations for preventing similar incidents. The alert updates a previous NIOSH Alert published in July 1985 [NIOSH 1985].

The recommendations in this alert should be followed by all employers, managers, supervisors, and workers in companies that use cranes or similar boomed vehicles. NIOSH requests that the following individuals and organizations bring this alert to the attention of workers who are at risk: editors of trade journals, safety and health officials, construction companies, unions, suppliers and manufacturers of building materials, crane manufacturers, electric utilities, and others who use cranes or boomed vehicles.

Workers are killed each year when cranes contact overhead power lines.

BACKGROUND

NTOF Data

Data from the NIOSH National Traumatic Occupational Fatalities (NTOF) Surveillance System indicate that electrocutions accounted for approximately 450 (7%) of the 6,400 work-related deaths from injury that occurred annually in the United States during the period 1980-89

[NIOSH 1993a]. Each year an average of 15 electrocutions were caused by contact between cranes or similar boomed vehicles and energized, overhead power lines. The actual number of workers who died from crane contact with energized power lines is higher than reported by NTOF because methods for collecting and reporting these data tend to underestimate the total number of deaths [NIOSH 1993a]. More than half of these crane-related electrocutions occurred in the construction industry.

FACE Data

From 1982 through 1994, NIOSH conducted 226 onsite investigations of work-related electrocutions under the Fatality Assessment and Control Evaluation (FACE) Program. Twenty-nine (13%) of these incidents (which resulted in 31 fatalities) involved crane contact with overhead power lines. Nearly half of the incidents occurred in the construction industry. Because the FACE investigations were conducted in only 16 states, these fatalities represent only a portion of the crane-related electrocutions during the period 1982-94.

OSHA Data

A study conducted by the Occupational Safety and Health Administration (OSHA) showed that 377 (65%) of 580 work-related electrocutions occurred in the construction industry during the period 1985-89 [OSHA 1990]. Nearly 30% (113) of these electrocutions involved cranes.

CURRENT STANDARDS

OSHA Regulations

Current OSHA regulations require employers to take precautions when cranes and boomed vehicles are operated near overhead power lines. Any overhead power line shall be considered energized unless the owner of the line or the electric utility company indicates that it has been de-energized and it is visibly grounded [29 CFR 1926.550 (a)(15)(vi)]. The OSHA regulations are summarized as follows:

Employers shall ensure that overhead power lines are de-energized or separated from the crane and its load by implementing one or more of the following procedures:

- De-energize and visibly ground electrical distribution and transmission lines [29 CFR 1910.333(c)(3); 29 CFR 1926.550(a)(15)]

- Use independent insulated barriers to prevent physical contact with the power lines [29 CFR 1910.333(c)(3); 29 CFR 1926. 550(a)(15)]

- Maintain minimum clearance between energized power lines and the crane and its load [29 CFR 1910.333(c)(3)(iii); 29 CFR 1926.550(a)(15)(i), (ii), (iii)].

Where it is difficult for the crane operator to maintain clearance by visual means, a person shall be designated to observe the clearance between the energized power lines and the crane and its load [29 CFR 1926.550(a)(15)(iv)].

The use of cage-type boom guards, insulating links, or proximity warning devices shall not alter the need to follow required precautions [29 CFR 1926.550(a)(15)(v)]. These devices are not a substitute for de-energizing and grounding lines or maintaining safe line clearances.

ANSI Standard

The American National Standards Institute (ANSI) has published a standard for mobile and locomotive cranes that includes operation near overhead power lines [ANSI 1994]. This consensus standard (B30.5-1994) contains guidelines for preventing contact between cranes and electrical energy. The standard addresses the following issues:

- Considering any overhead wire to be energized unless and until the person owning the line or the utility authorities verify that the line is not energized

- De-energizing power lines before work begins, erecting insulated barriers to prevent physical contact with the energized lines, or maintaining safe clearance between the energized lines and boomed equipment

- Limitations of cage-type boom guards, insulating links, and proximity warning devices

- Notifying line owners before work is performed near power lines

- Posting warnings on cranes cautioning the operators to maintain safe clearance between energized power lines and their equipment.

CSA Recommendations

The Construction Safety Association (CSA) of Ontario, Canada recommends safe work practices in addition to those addressed in the OSHA and ANSI standards [CSA 1982]. These recommendations include the following.

Work Practices

- Operate the crane at a slower-than-normal rate in the vicinity of power lines

- Exercise caution near long spans of overhead power lines, since wind can cause the power lines to sway laterally and reduce the clearance between the crane and the power line

- Mark safe routes where cranes must repeatedly travel beneath power lines

- Exercise caution when traveling over uneven ground that could cause the crane to weave or bob into power lines

- Keep all personnel well away from the crane whenever it is close to power lines

- Prohibit persons from touching the crane or its load until a signal person indicates that it is safe to do so.

The CSA recommendations also address the limitations of proximity warning devices, hook insulators, insulating boom guards, swing limit stops, nonconductive taglines, ground rods, and similar devices for protection against electrical hazards.

Procedures to Follow If Contact Occurs

To protect against electrical shock injury in the event of contact between a crane and an energized line, the CSA recommends the following:

- The crane operator should remain inside the cab

- All other personnel should keep away from the crane, ropes, and load, since the ground around the machine might be energized

- The crane operator should try to remove the crane from contact by moving it in the reverse direction from that which caused the contact

- If the crane cannot be moved away from contact, the operator should remain inside cab until the lines have been de-energized.

CASE REPORTS

The five cases presented here were investigated by the NIOSH FACE Program between March 1990 and March 1993.

Case No. 1 — One Death

On March 1, 1990, a 29-year-old worker was electrocuted when he pushed the crane cable on a 1-yard cement bucket into a 7,200-volt power line. The victim was a member of a crew that was constructing the back concrete wall of an underground water-holding tank at a sewage treatment plant. Before work on the tank began, the company safety director made sure that insulated line hoses were placed over sections of the power line near the jobsite and that a safe clearance zone was marked off for arriving cement trucks to use for loading their cement buckets.

After the wall was poured, the driver of the cement truck cleaned the loading chute on his truck with a water hose mounted on the truck. As he began to pull away, the crew supervisor yelled to him, asking if the crew could use his water hose to wash out the cement bucket suspended from the crane. The driver stopped the truck under the power line and the crane operator (not realizing that the truck had been moved) swung the boom to position the bucket behind the truck. The victim grasped the handle of the bucket door and pushed down to open it, bringing the crane cable into contact with the power line. The victim provided a path to ground and was electrocuted [NIOSH 1990b].

Case No. 2 — One Death

On August 11, 1990, a 33-year-old well driller was electrocuted when a metal pipe lifted by a truck-mounted crane contacted a 12,000-volt overhead power line. The victim and a coworker were repairing a submersible pump for a water well at a private residence. The well was located in a pasture with three parallel power lines overhead. One of the power lines passed directly over the well (32 feet above the ground). On the day of the incident, the victim positioned the truck-mounted crane beneath the power line. Using a handheld remote-control pendant, the victim fully extended the end of the boom 36 feet above the ground. The crane cable was attached to a 1-inch-diameter galvanized pipe that ran to the pump inside the well. As the victim raised the pipe, it contacted the power line directly above the well, energizing the crane and the handheld remote-control pendant. The victim provided a path to ground and was electrocuted [NIOSH 1990c].

Case No. 3 — One Death

On August 22, 1990, a 24-year-old foreman for a telecommunications company was electrocuted when he grabbed the door handle on a truck-mounted crane whose boom was in contact with a 7,200-volt overhead power line. The foreman and three other workers (a lineman, a cable splicer, and a laborer) were attempting to remove four poles that had supported a billboard. The poles stood 20 feet high and were buried 5 feet in the ground. They were located 15 feet away from (and parallel to) the power line. To remove the poles, the lineman positioned the crane directly under the power line. He controlled the crane boom while standing on the ground using rubber-coated hand controls mounted on the back of the truck. The poles were removed by hooking the crane boom cable around the middle of each pole and vertically pulling each pole out of the ground. While the workers were pulling out the third pole, the end of the boom contacted the overhead power line. The laborer (who was working in the back of the truck) noticed that the lineman was being shocked and was unable to let go of the hand control. The laborer kicked the lineman in the chest and the lineman fell unconscious to the ground. He revived without assistance about 3 minutes later with electrical burns to his left hand. However, the crane boom remained in contact with the power line, the truck tires ignited, and the truck began to burn. When the foreman noticed that the boom remained in contact with the power line, he tried to open one of the truck doors (presumably to move the truck). When his hand contacted the door handle, he provided a path to ground and was electrocuted [NIOSH 1990a].

Case No. 4 — One Death

On June 24, 1991, a 37-year-old construction laborer was electrocuted while pulling a wire rope attached to a crane cable toward a load. The choker was to be connected to a steel roof joist that was to be lifted 150 feet across the roof of a one-story school and set in place. The cab of the crane was positioned 11 feet 6 inches from a 7,200-volt power line. After a previous roof joist had been set in place, the crane operator swung the crane boom and cable back toward the victim, who grabbed the choker in his left hand. With his right hand, he held onto a steel rod that had been driven into the ground nearby. At this point, the momentum of the swinging crane apparently caused the crane cable to contact the power line. The electrical current passed across the victim's chest and through the steel rod to ground, causing his electrocution [NIOSH 1991].

Case No. 5 — Two Deaths

On March 31, 1993, a 20-year-old male truck driver and his 70-year-old male employer (the company president) were electrocuted when the boom of a truck-mounted crane contacted a 7,200-volt conductor of an overhead power line. The incident occurred while the driver was unloading concrete blocks at a residential construction site. The driver had backed the truck up the steeply sloped driveway under a power line at the site and was using the crane to unload a cube of concrete blocks. The company president and a masonry contractor watched as the driver operated the crane by a handheld remote-control unit. The driver was having difficulty unloading the blocks because the truck was parked at a steep angle. While all three men watched the blocks, the tip of the crane boom contacted a conductor of the overhead power line and completed a path to ground through the truck, the remote control unit, and the driver. The company president attempted to render assistance and apparently contacted the truck, completing a path to ground through his body. He died on the scene. The truck driver was airlifted to a nearby burn center where he later died as a result of electrical burns [NIOSH 1993b].

CONCLUSIONS

These case reports indicate that some crane operators, their employers and supervisors, and others who work around cranes may not be fully aware of the hazards of operating cranes near overhead power lines or may not implement the proper safety procedures for controlling these hazards.

RECOMMENDATIONS

NIOSH recommends that employers take the following measures to protect workers and operators of cranes and other boomed vehicles from contacting energized overhead power lines.

Comply with OSHA Regulations

Train workers to comply with current OSHA regulations. These regulations require workers and employers to consider all overhead power lines to be energized until (1) the owner of the lines or the electric utility indicates that they are not energized, and (2) they have been visibly grounded [29 CFR 910.333 (c)(3); 29 CFR 1926.550(a)(15)].

Employers shall ensure that overhead power lines are de-energized or separated from the crane and its load by implementing one or more of the following [29 CFR 1910.333(c)(3); 29 CFR 1926.550(a)(15)]:

- De-energize and visibly ground electrical distribution and transmission lines at the point of work

- Use insulated barriers that are not a part of the crane to prevent contact with the lines

- If the power lines are not de-energized, operate cranes in the area ONLY if a safe minimum clearance is maintained as follows:

  - At least 10 feet for lines rated 50 kilovolts or below

  - At least 10 feet plus 0.4 inch for each kilovolt above 50 kilovolts; or maintain twice the length of the line insulator (but never less than 10 feet).

  Where it is difficult for the crane operator to maintain safe clearance by visual means, designate a person to observe the clearance and to give immediate warning when the crane approaches the limits of safe clearance [29 CFR 1926.550(a) (15)(iv)].

  Do not use cage-type boom guards, insulating links, or proximity warning devices as a substitute for de-energizing and grounding lines or maintaining safe clearance [29 CFR 1926.550(a)(15)(v)].

Follow ANSI Guidelines

Train workers to follow ANSI guidelines for operating cranes near overhead power lines (ANSI Standard B30.5-1994, 5-3.4.5) [ANSI 1994]. These guidelines recommend posting signs at the operator's station and on the outside of the crane warning that electrocution may occur if workers do not maintain safe minimum clearance that equals or exceeds OSHA requirements as follows:

Power line voltage phase to phase (kV)	Minimum safe clearance (feet)
50 or below	10
Above 50 to 200	15
Above 200 to 350	20
Above 350 to 500	25
Above 500 to 750	35
Above 750 to 1,000	45

Notify Power Line Owners

Before beginning operations near electrical lines, notify the owners of the lines or their authorized representatives and provide them with all pertinent information: type of equipment (including length of boom) and date, time, and type of work involved. Request the cooperation of the owner to de-energize and ground the lines or to help provide insulated barriers. NIOSH encourages employers to consider de-energization (where possible) as the primary means of preventing injury from contact between cranes and power lines.

Develop Safety Programs

Develop and implement written safety programs to help workers recognize and control the hazards of crane contact with overhead power lines.

Evaluate Job Sites

Evaluate job sites before beginning work to determine the safest areas for material storage, the best placement for machinery during operations, and the size and type of machinery to be used.

Know the location and voltage of all overhead power lines at the jobsite before operating or working with any crane.

Research has shown that it is difficult to judge accurately the distance to an overhead object such as a power line [Middendorf 1978]. Therefore, NIOSH recommends that no other duties or responsibilities be assigned when workers are designated to observe clearance during crane movement or operation.

Evaluate Alternative Work Methods

Evaluate alternative work methods that do not require the use of cranes. For example, it may be possible to use concrete pumping trucks instead of crane-suspended buckets for placing concrete near overhead power lines. Alternative methods should be carefully evaluated to ensure that they do not introduce new hazards into the workplace.

Train Workers

Ensure that workers assigned to operate cranes and other boomed vehicles are specifically trained in safe operating procedures. Also ensure that workers are trained (1) to understand the limitations of such devices as boom guards, insulated lines, ground rods, nonconductive links, and proximity warning devices, and (2) to recognize that these devices are not substitutes for de-energizing and grounding lines or maintaining safe clearance. Workers should also be trained to recognize the hazards and use proper techniques when rescuing coworkers or recovering equipment in contact with electrical energy. CSA guidelines list techniques that can be used when equipment contacts energized power lines [CSA 1982] (see Current Standards in this Alert).

All employers and workers should be trained in cardiopulmonary resuscitation (CPR).

Call for Help

Ensure that workers are provided with a quick means of summoning assistance when an emergency occurs.

Develop Safer Equipment

Encourage the manufacturers of cranes and other boomed vehicles to consider developing truck-mounted cranes with electrically isolated crane control systems, such as those that use fiber optic conductors to transmit control signals.

ACKNOWLEDGMENTS

The principal contributor to this Alert is Paul H. Moore, Division of Safety Research. Please direct any comments, questions, or requests for additional information to the following:

> Director
> Division of Safety Research
> National Institute for Occupational Safety and Health
> 1095 Willowdale Road
> Morgantown, WV 26505-2888
> Telephone, (304) 285-5894; or call 1-800-35-NIOSH (1-800-356-4674).

We greatly appreciate your assistance in protecting the health of U.S. workers.

> Linda Rosenstock, M.D., M.P.H.
> Director, National Institute for
> Occupational Safety and Health
> Centers for Disease Control and Prevention

REFERENCES

1. ANSI [1994]. *American National Standard for Mobile and Locomotive Cranes.* New York, NY: American National Standards Institute, ANSI B30.5-1994.

2. CFR. *Code of Federal Regulations.* Washington, DC: U.S. Government Printing Office, Office of the Federal Register.

3. CSA (Construction Safety Association) [1982]. *Mobile Crane Manual.* Toronto, Ontario, Canada: Construction Safety Association of Ontario.

4. Middendorf, L., [1978]. Judging Clearance Distances near Overhead Power Lines. In: *Proceedings of the Human Factors Society*, 22nd Annual Meeting. Santa Monica, CA: Human Factors Society, Inc.

5. NIOSH [1985]. *NIOSH Alert: Request for Assistance in Preventing Electrocutions from Contact Between Cranes and Power Lines.* Cincinnati, OH: U.S. Department of Health and Human Services, Public Health Service, Centers for Disease Control, National Institute for Occupational Safety and Health, DHHS (NIOSH) Publication No. 85-111.

NIOSH [1990a]. *Foreman Electrocuted and Lineman Injured After Truck-mounted Crane Boom Contacts 7,200-volt Overhead Power Line in Virginia.* Morgantown, WV: U.S. Department of Health and Human Services, Public Health Service, Centers for Disease Control, National Institute for Occupational Safety and Health, Fatal Accident Circumstances and Epidemiology (FACE) Report No. 90-39.

NIOSH [1990b]. *Laborer Touching Suspended Cement Bucket Electrocuted When Crane Cable Contacts 7,200-volt Power Line in North Carolina.* Morgantown, WV: U.S. Department of Health and Human Services, Public Health Service, Centers for Disease Control, National Institute for Occupational Safety and Health, Fatal Accident Circumstances and Epidemiology (FACE) Report No. 90-29.

NIOSH [1990c]. *Well Driller Electrocuted When Pipe on Crane Cable Contacts 12,000-volt Overhead Power Line in Virginia.* Morgantown, WV: U.S. Department of Health and Human Services, Public Health Service, Centers for Disease Control, National Institute for Occupational Safety and Health, Fatal Accident Circumstances and Epidemiology (FACE) Report No. 90-38.

NIOSH [1991]. *Construction Laborer is Electrocuted When Crane Boom Contacts Overhead 7,200-volt Power Line in Kentucky.* Morgantown, WV: U.S. Department of Health and Human Services, Public Health Service, Centers for Disease Control, National Institute for Occupational Safety and Health, Fatal Accident Circumstances and Epidemiology (FACE) Report No. 91-21.

NIOSH [1993a]. *Fatal Injuries to Workers in the United States, 1980-1989: A Decade of Surveillance.* Cincinnati, OH: U.S. Department of Health and Human Services, Public Health Service, Centers for Disease Control and Prevention, National Institute for Occupational Safety and Health, DHHS (NIOSH) Publication No. 93-108.

NIOSH [1993b]. *Truck Driver and Company President Electrocuted After Crane Boom Contacts Power Line in West Virginia.* Morgantown, WV: U.S. Department of Health and Human Services, Public Health Service, Centers for Disease Control and Prevention, National Institute for Occupational Safety and Health, Fatality Assessment and Control Evaluation (FACE) Report No. 93-14.

OSHA [1990]. *Analysis of Construction Fatalities: The OSHA Database 1985-1989.* Washington, DC: U.S. Department of Labor, Occupational Safety and Health Administration.

Crane Alert — DHHS (NIOSH) Publication No. 95-108

1. *NIOSH Construction, Safety, and Health Research,* August 1997, page ii.

12. Jenkins EL, Kisner SM, Fosbroke DE, et al [1993]. Washington, D.C.: U.S. Government Printing Office. DHHS (NIOSH) publication 93-108.

13. Toscano G, Windau J [1996]. National Census of Fatal Occupational Injuries, 1995. Compensation and Working Conditions, September 1996: 34-45.

14. National Safety Council [1991]. *Accident Facts.* Chicago: National Safety Council.

17. NIOSH [1995]. NIOSH Alert: Request for assistance in preventing electrocutions of crane operators and crew members working near overhead power lines. Cincinnati, OH: U.S. Department of Health and Human Services, Centers for Disease Control and Prevention, National Institute for Occupational Safety and Health, Division of Safety Research. DHHS (NIOSH) Publication 95-108.

APPENDIX H

OSHA 1926.20 - GENERAL SAFETY AND HEALTH PROVISIONS

Standard Number: 1926.20

Standard Title: General safety and health provisions.

SubPart Number: C

SubPart Title: General Safety and Health Provisions

(a) Contractor Requirements

(a)(1)

Section 107 of the Act requires that it shall be a condition of each contract which is entered into under legislation subject to Reorganization Plan Number 14 of 1950 (64 Stat. 1267), as defined in 1926.12, and is for construction, alteration, and/or repair, including painting and decorating, that no contractor or subcontractor for any part of the contract work shall require any laborer or mechanic employed in the performance of the contract to work in surroundings or under working conditions which are unsanitary, hazardous, or dangerous to his health or safety.

(b) Accident Prevention Responsibilities.

(b)(1)

It shall be the responsibility of the employer to initiate and maintain such programs as may be necessary to comply with this part.

(b)(2)

Such programs shall provide for frequent and regular inspections of the job sites, materials, and equipment to be made by competent persons designated by the employers.

(b)(3)

The use of any machinery, tool, material, or equipment which is not in compliance with any applicable requirement of this part is prohibited. Such machine, tool, material, or equipment shall either be identified as unsafe by tagging or locking the controls to render them inoperable or shall be physically removed from its place of operation.

(b)(4)

The employer shall permit only those employees qualified by training or experience to operate equipment and machinery.

(c) Standards Application

The standards contained in this part shall apply with respect to employments performed in a workplace in a State, the District of Columbia, the Commonwealth of Puerto Rico, the Virgin Islands, American Samoa, Guam, Trust Territory of the Pacific Islands, Wake Island, Outer Continental Shelf lands defined in the Outer Continental Shelf Lands Act, Johnston Island, and the Canal Zone.

(d)(1)

If a particular standard is specifically applicable to a condition, practice, means, method, operation, or process, it shall prevail over any different general standard which might otherwise be applicable to the same condition, practice, means, method, operation, or process.

(d)(2)

On the other hand, any standard shall apply according to its terms to any employment and place of employment in any industry, even though particular standards are also prescribed for the industry to the extent that none of such particular standards applies.

(e)

In the event a standard protects on its face a class of persons larger than employees, the standard shall be applicable under this part only to employees and their employment and places of employment.

[44 FR 8577, Feb. 9, 1979; 44 FR 20940, Apr. 6, 1979, as amended at 58 FR 35078; June 30, 1993]

APPENDIX I

SLING SAFETY, OSHA 3072

This information booklet is intended to provide a generic, non-exhaustive overview of a particular standards-related topic. This publication does not itself alter or determine compliance responsibilities, which are set forth in OSHA standards themselves, and in the Occupational Safety and Health Act. Moreover, because interpretations and enforcement policy may change over time, for additional guidance on OSHA compliance requirements, the reader should consult current administrative interpretations and decisions by the Occupational Safety and Health Review Commission and the courts.

Introduction

The ability to handle materials—to move them form one location to another, whether during transit or at the worksite—is vital to all segments of industry. Materials must be moved, for example, for industry to manufacture, sell, and utilize products. In short, without material-handling equipment capability, industry would cease to exist.

To varying degrees, all employees in numerous workplaces take part in materials handling. Consequently, some employees are injured. In fact, the mishandling of materials is the single greatest cause of accidents and injuries in the workplace. Most of these accidents and injuries, as well as the pain and loss of salary and productivity that often result, can be readily avoided. Whenever possible, mechanical means should be used to move materials to avoid employee injuries such as muscle pulls, strains, and sprains. In addition, many loads are too heavy and/or bulky to be safely moved manually. Various types of equipment, therefore, have been designed specifically to aid in the movement of materials: cranes, derricks, hoists, powered industrial trucks, and conveyors. Because cranes, derricks, and hoists rely upon slings to hold their suspended loads, slings are the most commonly uses materials-handling apparatus. This booklet offers information on the proper selection, maintenance, and use of slings.

This booklet is designed to assist sling operators to understand and comply with OSHA's regulations on slings, published on June 27, 1995, in the Federal Register under Title 29 Code of Federal Regulations, Part 1910.184.

All employers should be aware that there are certain states (and similar jurisdictions) which operate their own programs under agreement with the U.S. Department of Labor, pursuant to section 18 of the Act. The programs in these jurisdictions may differ in some detail from the federal program. (See list of States with Approved Plans in Appendix G.)

Importance of the Operator

The operator must exercise intelligence, care, and common sense when selecting and using slings. Slings must be selected in accordance with their intended use, based upon the size and type of load, and the environmental conditions of the workplace. All slings must be visually inspected before use to ensure their effectiveness.

A well-trained operator can prolong the service life of equipment and reduce costs by avoiding the potentially hazardous effects of overloading equipment, operating it at excessive speeds, taking up slacks with a sudden jerk, and suddenly accelerating or decelerating equipment. The operator can look for causes and seek corrections whenever a danger exists. He or she should cooperate with coworkers and supervisors and become a leader in carrying out safety measures B not merely for the good of the equipment and the production schedule but, more importantly, for the safety of everyone concerned.

Sling Types

The dominant characteristics of a sling are determined by the components of that sling. For example, the strengths and weaknesses of a wire rope sling are essentially the same as the strengths and weaknesses of the wire rope of which it is made.

Slings are generally one of six types: chain, wire rope, metal mesh, natural fiber rope, synthetic rope, or synthetic web. In general, use and inspection procedures tend to place these slings into three groups: chain, wire rope and mesh, and fiber rope web. Each type has its own particular advantages and disadvantages. Factors to consider when choosing the best sling for the job include the size, weight, shape, temperature, and sensitivity of the material to be moved, as well as the environmental conditions under which the sling will be used.

Chains

Chains are commonly used because of their strength and ability to adapt to the shape of the load. Care should be taken, however, when using alloy chain slings because sudden shocks will damage them. Misuse of chain slings could damage the sling, resulting in sling failure and possibly injury to an employee.

Chain slings are best for lifting very hot metals. They can be heated to temperatures of up to 1000° F (538° C); however, when alloy chain slings are consistently exposed to service temperatures in excess of 600° F(316° C), operators must reduce the working load limits in accordance with the manufacturer's recommendations.

All sling types must be visually inspected prior to use. When inspecting alloy steel chain slings, pay special attention to any stretching, wear in excess of the allowances made by the manufacturer, and nicks and gouges. These signs indicate that the sling may be unsafe and they must be removed from service.

Wire Ropes

A second type of sling is made of wire rope. Wire rope is composed of individual wires that have been twisted to form strands. Strands are then twisted to form a wire rope. When wire rope has a fiber core, it is usually more flexible but is less resistant to environmental damage. Conversely, a core that is made of wire rope strands tends to have greater strength and is more resistant to heat damage.

Wire rope may be further defined by the "lay." The lay of a wire rope describes the direction the wires and strands are twisted during the construction of the rope. Most wire rope is right lay, regular lay—which means that the strands pass from left to right across the rope and the wires in the rope are laid opposite in direction to the lay of the strands. This type of rope has the widest range of applications.

Lang lay (where the wires are twisted in the same direction as the strands) is recommended for many excavating, construction, and mining applications, including draglines, hoist lines, dredgelines, and other similar lines.

Lang lay ropes are more flexible and have greater wearing surface per wire than regular lay ropes. In addition, since the outside wires in lang lay rope lie at an angle to the rope axis, internal stress due to bending over sheaves and drums is reduced causing lang lay ropes to be more resistant to bending fatigue.

A left lay rope is one in which the strands form a left-hand helix similar to the threads of a left-hand screw thread. Left lay rope has its greatest usage in oil fields on rod and tubing lines, blast hole rigs, and spudders where rotation of right lay would loosen couplings. The rotation of a left lay rope tightens a standard coupling.

Wire Rope Sling Selection. When selecting a wire rope sling to give the best service, there are four characteristics to consider: strength, ability to bend without distortion, ability to withstand abrasive wear, and ability to withstand abuse.

1.   *Strength*: The strength of a wire rope is a function of its size, grade, and construction. It must be sufficient to accommodate the applied maximum load. The maximum load limit is determined by means of an appropriate multiplier. This multiplier is the number by which the ultimate strength of a wire rope is divided to determine the working load limit. Thus, a wire rope sling with a strength of 10,000 pounds (4,545 kilograms) and a total working load of 2,000 pounds (909 kilograms) has a design factor (multiplier) of 5. New wire rope slings have a design factor of 5. As a sling suffers from the rigors of continued service, however, both the design factor and the sling's ultimate strength are proportionately reduced. If a sling is loaded beyond its ultimate strength, it will fail. So, older slings must be more rigorously inspected to ensure that rope conditions adversely affecting the strength of the sling are considered in determining if a wire rope sling should be allowed to continue in service.

2.   *Fatigue (Bending without Failure)*: wire rope must have the ability to withstand repeated bending without the wires failing from fatigue. Fatigue failure of the wires in a wire rope is the result of the development of small cracks from repeated applications of bending loads. It occurs when ropes make small radius bends. The best means of preventing fatigue failure of wire rope slings is to use blocking or padding to increase the radius of bend.

3.   *Abrasive Wear*: The ability of a wire rope to withstand abrasion is determined by the size, number of wires, and construction of the rope. Smaller wires bend more readily and therefore offer greater flexibility but are less able to withstand abrasive wear. Conversely, the larger wires of less flexible ropes are better able to withstand abrasion than are the smaller wires of more flexible ropes.

4.   *Abuse*: All other factors being equal, misuse or abuse of wire rope will cause a wire rope sling to become unsafe long before any other factor. Abusing a wire rope sling can cause serious structural damage to the wire rope, such as kinking or bird caging, which reduces the strength of the wire rope. (In bird caging, the wire rope strands are forcibly untwisted and become spread outward.) So, to prolong the life of the sling and protect the lives of

employees, the manufacturer's suggestion for safe and proper use of wire rope slings must be strictly adhered to.

Wire Rope Life. Many operating conditions affect wire rope life. They are bending, stresses, loading conditions, speed of load application jerking, abrasion, corrosion, sling design, materials handled, environmental conditions, and history of previous usage.

In addition to the above operating conditions, the weight, size, and shape of the loads to be handled also affect the service life of a wire rope sling. Flexibility is also a factor. Generally, more flexible ropes are selected when smaller radius bending is required. Less flexible ropes should be used when the rope must move through or over abrasive materials.

Wire Rope Sling Inspection. Wire rope slings must be visually inspected before each day's use. The operator should check the twists or lay of the sling. If ten randomly distributed wires in one lay are broken, or five wires in one strand of a rope lay are damaged, the sling must not be used. It is not sufficient, however, to check only the condition of the wire rope. End fittings and other components should also be inspected for any damage that could make the sling unsafe to use.

To ensure safe sling usage between scheduled inspections, all workers should participate in a safety awareness program. Each operator should keep a close watch on those stings he or she is using. If any accident involving the movement of materials occurs, the operator should immediately shut down the equipment and report the accident to a supervisor. The cause of the accident should be determined and corrected before resuming operations.

Field Lubrication. Although every rope sling is lubricated when manufactured, it must also be lubricated "in the field" to increase the sling's useful service life. There is no set rule on how much or how often this should be done. It depends on the conditions under which the sling is used. The heavier the loads, the greater the number of bends, or the more adverse the conditions under which the sling operates, the more frequently lubrication is required.

Storage. Wire rope slings should be stored in a well-ventilated, dry building or shed. To avoid corrosion and rust, never store wire rope slings on the ground or allow them to be continuously exposed to the elements. And, if it is necessary to store wire rope slings outside. make sure that they are set off the ground and protected.

> *Note: Using the sling several times a week, even with light loads, is a good practice. Records show that frequently or continuously used slings give useful service far longer than idle ones.*

Discarding Slings. Wire rope slings can provide a margin of safety by showing early signs of failure. The following factors indicate when a wire sling needs to be discarded:

- Severe corrosion

- Localized wear (shiny worn spots) on the outside

- A one-third reduction in outer wire diameter

- Damage or displacement of end—fittings—hooks, rings, links, or collars—by overload or misapplication

- Distortion, kinking, bird caging, or other evidence of damage to the wire rope structure

- Excessive broken wires.

Fiber Rope and Synthetic Web

Fiber rope and synthetic web slings are used primarily for temporary work, such as construction and painting jobs, and in marine operations. They also are the best choice for use on expensive loads, highly finished parts, fragile parts, and delicate equipment.

Fiber Rope Slings. Fiber rope deteriorates on contact with acids and caustics. Fiber ropes slings, therefore, must not be used around these substances unless the manufacturer recommends them for that use.

When inspecting a fiber rope sling, look first at its surface. Look for cuts, gouges, or worn surface areas; dry, brittle scorched, or discolored fibers; or melting or charring of any part of the sling. If any of these conditions are found, the supervisor must be notified and a determination made regarding the safety of the sling. If the sling is found to be unsafe, it must be discarded.

Next check the sling's interior. It should be as clean as when the rope was new. A buildup of powder like sawdust on the inside of the fiber rope indicates excessive internal wear and that the sling is unsafe.

Finally, scratch the fibers with a fingernail. If the fibers separate easily, the fiber sling has suffered some kind of chemical damage and must be discarded.

Synthetic Rope and Web Slings. The most commonly used synthetic web slings are made of nylon, polypropylene, and polyester. They have the following properties in common:

- *Convenience*: can conform to any shape

- *Economy*: have a low initial cost plus a long service life

- *Load protection*: will not mar, deface, or scratch highly polished or delicate surfaces

- *Long life*: are unaffected by mildew, rot, or bacteria; resist some chemical action; and have excellent abrasion resistance

- *Safety*: will adjust to the load contour and hold it with a tight, non-slip grip

- *Shock absorbency*: an absorb heavy shocks without damage

- *Strength*: can handle a load of up to 300,000 pounds (136,363 kilograms)

- *Temperature resistance*: are unaffected by temperatures up to 180° F (82.2° C).

Because each synthetic material has unique properties, it should be used according to the manufacturer's instructions, especially when dealing with chemically active environments.

Possible Defects. Synthetic web slings must be removed from service if any of the following defects exist:

- Acid or caustic burns

- Melting or charring of any part of the surface

- Snags, punctures, tears, or cuts

- Broken or worn stitches

- Wear or elongation exceeding the amount recommended by the manufacturer

- Distortion of fittings.

Safe Lifting Practices

Now that the sling has been selected (based upon the characteristics of the load and the environmental conditions surrounding the lift) and inspected prior to use, the next step is learning how to use it safely. There are four primary factors to consider when safely lifting a load. They are (1) the size, weight, and center of gravity of the load; (2) the number of legs and the angle the sling makes with the horizontal line; (3) the rated capacity of the sling; and (4) the history of the care and usage of the sling.

Size, Weight, and Center of Gravity of the Load

The center of gravity of an object is that point at which the entire weight may be considered as concentrated. To make a level lift, the crane hook must be directly above this point. While slight variations are usually permissible, if the crane hook is too far to one side of the center of gravity, dangerous tilting will result causing unequal stresses in the different sling legs. This imbalance must be compensated for at once.

Number of Legs and Angle with the Horizontal

As the angle formed by the sling leg and the horizontal line decreases, the rated capacity of the sling also decreases. In other words, the smaller the angle between the sling leg and the horizontal, the greater the stress on the sling leg and the smaller (lighter) the load the sling can safely support. Larger (heavier) loads can be safely moved if the weight of the load is distributed among more sling legs.

The rated capacity of a sling varies depending upon the type of sling, the size of the sling, and the type of hitch. Operators must know the capacity of the sling. Charts or tables that contain this information generally are available from sling manufacturers. The values given are for new slings. Older slings must be used with additional caution. Under no circumstances shall a sling's rated capacity be exceeded.

History of Care and Usage

The mishandling and misuse of slings are the leading cause of sling-related accidents. The majority of injuries and accidents, however, can be avoided by becoming familiar with the essentials of proper sling care and use.

Proper care and use are essential for maximum service and safety. Slings must be protected with cover saddles, burlap padding, or wood blocking, as well as from unsafe lifting procedures such as overloading to prevent sharp bends and cutting edges.

Before making a lift, check to be certain that the sling is properly secured around the load and that the weight and balance of the load have been accurately determined. If the load is on the ground, do *not* allow the load to drag along the ground. This could damage the sling. If the load is already resting on the sling, ensure that there is no sling damage prior to making the lift.

Next, position the hook directly over the load and seat the sling squarely within the hook bowl. This gives the operator maximum lifting efficiency without bending the hook or overstressing the sling.

Wire rope slings also are subject to damage resulting from contact with sharp edges of the loads being lifted. These edges can be blocked or padded to minimize damage to the sling.

After the sling is properly attached to the load, there are a number of good lifting techniques common to all slings. First, make sure that the load is not lagged, clamped, or bolted to the floor. Second, guard against shock loading by taking up the slack in the sling slowly. Apply power cautiously to prevent jerking at the beginning of the lift, and slowly accelerate or decelerate. Third, check the tension on the sling. Raise the load a few inches, stop, and check for proper balance and that all items are clear of the path of travel. Never allow anyone to ride on the hood or load. Fourth, keep all personnel clear while the load is being raised, moved, or lowered. Crane or hoist operators should watch the load at all times when it is in motion. Finally, obey the following "nevers." Never allow more than one person to control a lift or give signals to a crane or hoist operator except to warn of a hazardous situation. Never raise the load more than necessary. Never leave the load suspended in the air, and never work under a suspended load or allow anyone else to.

Once the lift has been completed, clean the sling, check it for damage, and store it in a clean, dry, airy place. It is best to hang it on a rack or wall.

Remember, damaged slings cannot lift as much weight as new or older well-cared for slings. Proper and safe use and storage of slings will increase their service life.

Maintenance of Slings

Chains

Chain slings must be cleaned prior to each inspection, as dirt or oil may hide damage. The operator must be certain to inspect the total length of the sling, periodically looking for stretching, binding, wear, or nicks and gouges. If a sling has stretched so that it is now more than three percent longer than when it was new, it is unsafe and must be discarded.

Binding is the term used to describe the condition that exists when a sling has become deformed to the extent that its individual links cannot move within each other freely. It indicates that the sling is unsafe. Generally, wear occurs on the load-bearing inside ends of the links. Flushing links together so that the inside surface becomes clearly visible is the best way to check for this type of wear. Wear may also occur, however, on the outside of links when the chain is dragged along abrasive surfaces or pulled out from under heavy loads. Either type of wear weakens slings and makes accidents more likely.

Heavy nicks or gouges must be filed smooth, measured with calipers, and then compared with the manufacturer's minimum allowable safe dimensions. When in doubt, or in borderline situations, do not use the sling. In addition, never attempt to repair the welded components on a sling. If the sling needs repair of this nature, the supervisor must be notified.

Wire Rope

Wire rope slings, like chain slings, must be cleaned prior to each inspection because they are subject to damage hidden by dirt or oil. In addition, they must be lubricated according to manufacturer's instructions. Lubrication prevents or reduces corrosion and wear due to friction and abrasion. Before applying any lubricant, however, the sling user should make certain that the sling is dry. Applying lubricant to a wet or damp sling traps moisture against the metal and hastens corrosion.

Corrosion deteriorates wire rope. It may be indicated by pitting but it is sometimes hard to detect. If a wire rope sling shows any sign of significant deterioration, that sling must be removed until it can be examined by a person who is qualified to determine the extent of the damage.

By following the above guidelines to proper sling use and maintenance, and by the avoidance of kinking, it is possible to greatly extend the useful service life of a wire rope sling.

Fiber and Synthetic Ropes

Fiber ropes and synthetic webs are generally discarded rather than serviced or repaired. Operators must always follow the manufacturer's recommendations.

Summary

There are good practices to follow to protect yourself while using slings to move materials.

- First, accept the responsibility for your own actions. Become a competent and careful employee. Your own life or that of your fellow workers or others may depend on it.

- Second, learn as much as you can about the materials with which you will be working. Slings come in many different types, one of which is right for your purpose.

- Third, analyze the load to be moved in terms of size, weight, shape, temperature, and sensitivity, then choose the sling which best meets those needs.

- Fourth, always inspect all the equipment before and after a move. Always be sure to give equipment whatever "in service" maintenance it may need.

- Fifth, use safe lifting practices. Use the proper lifting technique for the type of sling and the type of load.

APPENDIX J

OSHA 3100 - CRANE OR DERRICK
SUSPENDED PERSONNEL PLATFORMS

U.S. Department of Labor
Occupational Safety and Health Administration
OSHA 3100
1993 (Revised)

Introduction

Using cranes or derricks to hoist personnel poses a significant risk to employees being lifted. To help prevent employee injury or death, the Occupational Safety and Health Administration (OSHA) regulation, Title 29 CFR 1926.550, limits the use of personnel hoisting in the construction industry and prescribes the proper safety measures for these operations.

Personnel platforms suspended from the load line and used in construction are covered by 29 CFR 1926.550(g). In addition, there is no specific provision for suspended personnel platforms in Part 1910. The governing provision, therefore, is general provision 1910.180(h)(3)(v), which prohibits hoisting, lowering, swinging, or traveling while anyone is on the load or hook. OSHA has determined, however, that when the use of a conventional means of access to any elevated work site would be impossible or more hazardous, a violation of 1910.180(h)(3)(v) will be treated as "de minimis" if the employer has complied with the provisions set forth in 1926.550(g)(3), (4), (5), (6), (7), and (8).

The OSHA rule for hoisting personnel is written in performance-oriented language that allows employers flexibility in deciding how to provide the best protection for their employees against the hazards associated with hoisting operations, and how to bring their work sites into compliance with the requirements of the standard.

Cranes and derricks used to hoist personnel must be placed on a firm foundation and the crane or derrick must be uniformly level within 1 percent of level grade.

The crane operator must always be at the controls when the crane is running and the personnel platform is occupied. The crane operator must also have full control over the movement of the personnel platform. Any movement of the personnel platform must be performed slowly and cautiously without any sudden jerking of the crane, derrick, or platform. Wire rope used for personnel lifting must have a minimum safety factor of seven. (This means it must be capable of supporting seven times the maximum intended load.) Rotation resistant rope must have a safety factor of ten.

When the occupied personnel platform is in a stationary position, all brakes and locking devices on the crane or derrick must be set.

The combined weight of the loaded personnel platform and its rigging must not exceed 50 percent of the rated capacity of the crane or derrick for the radius and configuration of the crane or derrick.

INSTRUMENTS AND COMPONENTS

Cranes and derricks with variable angle booms must have a boom angle indicator that is visible to the operator. Cranes with telescoping booms must be equipped with a device to clearly indicate the boom's length, or an accurate determination of the load radius to be used during the lift must be made prior to hoisting personnel. Cranes and derricks also must be equipped with (1) an anti-two blocking device that prevents contact between the load block, the overhaul ball, and the boom tip, or (2) a two-block damage-prevention feature that deactivates the hoisting action before damage occurs.

PERSONNEL PLATFORMS

Platforms used for lifting personnel must be designed with a minium safety factor of five and by a qualified person competent in structural design. The suspension system must be designed to minimize tipping due to personnel movement on the platform.

Each personnel platform must be provided with a standard guardrail system that is enclosed from the toeboard to the mid-rail to keep tools, materials, and equipment from falling on employees below, The platform also must have an inside grab rail, adequate headroom for employees, and a plate or other permanent marking that clearly indicates the platform's weight and rated load capacity or maximum intended load. When personnel are exposed to falling objects, overhead protection on the platform and the use of hardhats are required.

An access gate, if provided, must not swing outward during hoisting and must have a restraining device to prevent accidental opening.

All rough edges on the platform must be ground smooth to prevent injuries to employees.

All welding on the personnel platform and its components must be performed by a qualified welder who is familiar with weld grades, types, and materials specified in the platform design.

LOADING

The personnel platform must not be loaded in excess of its rated load capacity or its minimum intended load. Only personnel instructed in the requirements of the standard and the task to be performed—along with their tools, equipment, and materials needed for the job—are allowed on the platform. Materials and tools must be secured and evenly distributed to balance the load while the platform is in motion.

RIGGING

When a wire rope bridle is used to connect the platform to the load line, the bridle legs must be connected to a master link or shackle so that the load is evenly positioned among the bridle legs. Bridles and associated rigging for attaching the personnel platform to the hoist line must not be used for any purpose.

Attachment assemblies such as hooks must be closed and locked to eliminate the hook throat opening; an alloy anchor-type pin may be used as an alternative. "Mousing" (wrapping wire around a hook to cover the hook opening) is not permitted.

INSPECTING AND TESTING

A trial lift of the unoccupied personnel platform must be made before any employees are allowed to be hoisted. During the trial lift, the personnel platform must be loaded at least to its anticipated lift weight. The lift must start at ground level or at the location where employees will enter the platform and proceed to each location where the personnel platform is to be hoisted and positioned. **The trial lift must be performed immediately prior to placing personnel on the platform.**

The crane or derrick operator must check all systems, controls, and safety devices to ensure the following:

- ▶ They are functioning properly

- ▶ There are no interferences

- ▶ All boom or hoisting configurations necessary to reach work locations will allow the operator to remain within the 50-percent load limit of the hoist's rated capacity.

If a crane or derrick is moved to a new location or returned to a previously used one, the trial lift must be repeated before hoisting personnel.

After the trial lift, the personnel platform must be hoisted a few inches and inspected to ensure that it remains secured and is properly balanced.

Before employees are hoisted, a check must be made to ensure the following:

- ▶ Hoist ropes are free of kinks

- ▶ Multiple lines are not twisted around each other

- ▶ The primary attachment is centered over the platform

- ▶ There is no slack in the wire rope

- ▶ All ropes are properly seated on drums and in sheaves.

Immediately after the trial lift, a thorough visual inspection of the crane or derrick, the personnel platform, and the crane or derrick base support or ground must be conducted by a competent person to determine if the test lift exposed any defects, or produced any adverse effects on component or structure. Any defects found during inspections must be corrected before hoisting personnel.

When initially brought to the job site and after any repair or modification, and prior to hoisting personnel, the platform and rigging must be proof-tested to 125 percent of the platform's rated capacity. This is achieved by holding the loaded platform—with the load evenly distributed—in a suspended position for five minutes. Then a competent person must inspect the platform and rigging for defects. If any problems are detected, they must be corrected and another proof test must be conducted. Personnel hoisting must not be conducted until the proof-testing requirements are satisfied.

PRE-LIFT MEETING

The employer must hold a meeting with all employees involved in personnel hoisting operations (crane or derrick operator, signal person(s), employees to be lifted, and the person responsible for the hoisting operation) to review the OSHA requirements and the procedures to be followed before any lift operations are performed.

This meeting must be held before the trial lift at each new work site and must be repeated for any employees newly assigned to the operation.

SAFE WORK PRACTICES

Employees, too, can contribute to safe personnel hoisting operations and help reduce the number of accidents and injuries associated with personnel hoisting operations. Employees must follow these safe work practices:

- ▸ Use tag lines unless their use creates an unsafe condition

- ▸ Keep all body parts inside the platform during raising, lowering, and positioning

- ▸ Make sure a platform is secured to the structure where work is to be performed before entering or exiting it, unless such securing would create an unsafe condition

- ▸ Wear a body belt or body harness system with a lanyard. The lanyard must be attached to the lower load block or overhaul ball or to a structural member within the personnel platform. If the hoisting operation is performed over water, the requirements 29 CFR 1926.106—Working over or near water—must apply

- ▸ Stay in view of, or in direct communication with, the operator or signal person.

Crane and derrick operators must follow these safe work practices:

- ▸ Never leave the crane or derrick controls when the engine is running or when the platform is occupied

- ▸ Stop all hoisting operations if there are indications of any dangerous weather conditions or other impending danger

- ▸ Do not make any lifts on another load line of a crane or derrick that is being used to hoist personnel.

Movement of Cranes

Personnel hoisting is prohibited while the crane is traveling except when the employer demonstrates that this is the least hazardous way to accomplish this task or when portal, tower, or locomotive cranes are used.

When cranes are moving while hoisting personnel, the following rules apply:

- ▸ Travel must be restricted to a fixed track or runway

- ▸ Travel also must be limited to the radius of the boom during the lift

- ▸ The boom must be parallel to the direction of the travel.

▸ There must be a complete trial run before employees occupy the platform.

▸ If the crane has rubber tires, the condition and air pressure of the tires must be checked and the chart capacity for lifts must be applied to remain under the 50-percent limit of the hoist's rated capacity. Outriggers may be partially retracted as necessary for travel.

Compliance with the common-sense requirements of the OSHA standard and *the determination that no other safe method is available* should greatly reduce or eliminate the injuries and accidents that occur too frequently during personnel hoisting operations.

APPENDIX K

NATIONAL CRANE OPERATOR CERTIFICATION PROGRAM SELECTED FOR STASTICAL STUDY

FOR IMMEDIATE RELEASE

CONTACT: Graham J. Brent
National Commission for the Certification of Crane Operators
Phone: (703) 560-2391
Fax: (703) 560-2392

Fairfax, Virginia, May 26, 1998 — The National Commission for the Certification of Crane Operators (CCO) has been selected by the National Institute of Occupational Safety & Health (NIOSH) for a study of the effectiveness of crane operator certification.

The two-year study will examine incidents by type of crane and type of error (e.g. operator error, power line contact, load handling, etc.) for each incident. It will test the hypothesis that there should be a decrease in the number of skills-related incidents after certification is put in place.

Dr. Nancy Nelson of NIOSH's Division of Safety Research, based in Morgantown, West Virginia, said that the CCO program represented an excellent opportunity for an "intervention-type" study comparing incident rates of a company before and after it implemented a CCO certification policy.

A similar NIOSH study of forklifts revealed a dramatic decline in incidents following training, a conclusion which led to the development of OSHA's forklift training rule. NIOSH conducts much of the research for OSHA.

The National Commission for the Certification of Crane Operators (CCO) was formed in January 1995 to develop effective performance standards for safe crane operation to assist all segments of general industry and construction. Since that time, more than 5,000 operators have been tested through 170 separate test administrations in 30 states.

APPENDIX L

CRITICAL LIFT PLAN

This lift plan is required on all lifts that have been designated as critical. This designation could be due to the load requiring exceptional care in handling because of size, weight, close-tolerance installation, high susceptibility to damage, or other unusual factors.

A pre-lift meeting must be conducted involving all participants to review the lift plan and resolve any questions.

Person completing the lift plan: _____ Date: _____

1.   Item(s) to be lifted: _____

_____

Scheduled lift date: _____

Location of lift: _____

Describe lift, including weight, key dimensions and center of gravity:

_____

_____

Specify if a trial lift is required: No ☐   Yes ☐

Does the lift consist of a hazardous or contaminated material?   No ☐   Yes ☐
If Yes, identify:

_____

2. Hoisting/Rigging equipment to be used:

Hoisting Equipment   Lifting Capacity

_____

_____

_____

Rigging and Below the Hook Hardware

Rigging/Hardware   Lifting Capacity

_____

_____

_____

3. List any required permits:

_____

_____

_____

4. Job Hazard Analysis/Job Safety Analysis Required: ☐ Yes ☐ No

If Yes, List Document #   _____

5. Safety Considerations:

▸ List required personal protective equipent: _____ |

▸ Does the work require any area to be flagged and/or blocked?   ☐ Yes ☐ No

     If yes, appropriate people notified:

Are power lines in crane lift zone?   ☐ Yes Distance _____   ☐ No

6. Special instructions: ☐ Yes (list below)   ☐ No

_____

_____

7. Pre-lift inspection to be completed before beginning the lift:

▸ Total weight of lift including rigging and below the hook hardware:

▸ Pre-use inspection of hoisting equipment:

▸ Pre-use inspection of rigging and below the hook hardware:

▸ Dimensions, center of gravity, and arrangements in accordance with **attached** rigging sketches

▸ Mobile crane set up with outriggers fully extended, pads on solid footing, tires clear of ground and crane level

▸ Trial lift complete (when specified in Section 1)

8.   The following employees will be participating in the lift, have attended the pre-lift meeting, reviewed the lift plan, and understand the procedure and equipment to be used:

Employees signatures:

1. _____   2. _____

3. _____   4. _____

5. _____   6. _____

7. _____   8. _____

Supervisor/job foreman signature: _____

Date: _____

APPENDIX M

CHECKLISTS

1. OVERHEAD AND GANTRY CRANES MONTHLY INSPECTION CHECKLIST

2. OVERHEAD AND GANTRY CRANES PERIODIC INSPECTION CHECKLIST

3. MOBILE CRANE FREQUENT INSPECTION CHECKLIST

4. MOBILE CRANE PERIODIC INSPECTION CHECKLIST

5. ASSESSMENT CHECKLIST: USE OF AN OVERHEAD BRIDGE CRANE AS A WORK PLATFORM

6. SUSPENDED PERSONNEL PLATFORM AUTHORIZATION FORM

CHECKLIST # 1

OVERHEAD AND GANTRY CRANES
MONTHLY INSPECTION CHECKLIST

CRANE OPERATION	COMMENTS
Operate trolley in both directions	
Operate bridge in both directions	
Test limit switches (with no load on the hook)	
Operate all controls	
CRANE INSPECTION	
Check for uniform reeving of rope	
Inspect hooks for cracks, deformation, chemical damage, and heat damage. Inspect hook throat latches for noticeable damage or wear. Record hook serial number:	
Inspect chains for nicks, gouges, distortion, wear, and corrosion	
Inspect running/hoisting ropes for significant wear, broken wires, kinking, crushing, birdcaging, corrosion, and sufficient lubrication	
Inspect electrical connections for damage	
Inspect air or hydraulic systems for leakage	
Check for operational fire extinguishers in cab-operated cranes	
Inspected by:	
Crane #:	Date of Inspection:

Any deficiencies found that could reduce the crane's capacity or adversely affect its safety must be corrected before the crane is placed back in operation.

CHECKLIST # 2

OVERHEAD AND GANTRY CRANES
PERIODIC INSPECTION CHECKLIST

CRANE OPERATION	COMMENTS
Operate trolley in both directions	
Operate bridge in both directions	
Test limit switches	
Check for uniform reeving of rope	
Operate all controls	
CRANE INSPECTION	
Inspect hooks for cracks, deformation, chemical damage, latch engagement (if provided), and heat damage. Hook serial number:	
Inspect chains for nicks, gouges, distortion, wear, and corrosion	
Inspect running ropes for significant wear, broken wires, kinking, crushing, birdcaging, corrosion, and sufficient lubrication	
Inspect electrical connections for damage	
Inspect air or hydraulic systems for leakage	
Inspected by:	
Crane#:	Date of Inspection:

Any deficiencies found that could reduce the crane's capacity or adversely affect its safety must be corrected before the crane is placed back in operation.

CHECKLIST # 3

**MOBILE CRANE
FREQUENT INSPECTION CHECKLIST**

CRANE INSPECTION	COMMENTS
Crankcase oil	
Coolant	
Hydraulic oil	
Gauges	
Service/parking brake	
Swing brake	
Air pressure	
All required charts/signs posted	
Tire inflation pressure	
Anti-two-block	
Load moment indicator	
Boom length indicator	
Boom angle indicators	
Inspect hooks for cracks, deformation, chemical damage, latch engagement (if provided), and heat damage. Hook serial number:	
Inspect chains for nicks, gouges, distortion, wear, and corrosion	
Inspect running ropes for significant wear, broken wires, kinking, crushing, birdcaging, corrosion, and sufficient lubrication. Ropes reeving in compliance with crane manufacturers specifications.	
Inspect electrical connections for damage	
Inspect air or hydraulic hoses/systems for leakage	
Inspected by:	
Crane#:	Date of Inspection:

Any deficiencies found that could reduce the crane's capacity or adversely affect its safety must be corrected before the crane is placed back in operation.

CHECKLIST # 4

MOBILE CRANE
PERIODIC INSPECTION CHECKLIST

CRANE OPERATION	COMMENTS
Check for deformed, cracked or corroded members, cracked welds, loose bolts, pins or rivets in the crane structure.	
Check outriggers: Horizontal cylinders and hoses Vertical cylinders and hoses Beams for straightness and cracked welds Boxes for cracked welds Floats or outrigger pads for cracks Pins and keepers Holding valves	
Check Boom: Straightness in tip, intermediate, dead, and base section. Wear plates Point sheaves Anchoring of rope Rope retainer Boom extension hose and reel Hinge pin Limit Switch; Warning device	
Jibs	
Lattice Booms; Cross members Lacing Cords	
Load Blocks; Sheaves, sheave bearings Plates Pins Swivel Hooks (crack detecting inspection) Hoist rope retainer Anchoring of rope Hook safety latch Proper rope reeving Load block and call labeled with capacity and weight	

Winch motor - Mounting bolts Drum condition Rope anchoring Rope spooling Brakes Drum speed indicator Hydraulic hose and fittings	
Counterweight - Slides Warning signs Mounting bolts	
Power plants - Gasoline Diesel Electric	
Brake and clutch system parts - Linings Pawls Ratchets	
Load, boom angle, and other operating aids - Radius versus boom angle Boom length indicator	
Radiators/oil coolers - Leakage Improper performance Blockage of air passage	
Chain drive sprockets - Excessive wear Excessive chain stretch	
Check travel steering, braking, and locking devices - for malfunctioning	
Inspect for rust on piston rods and control valves when crane has been idle	
Hydraulic and pneumatic hose, fittings, and tubing inspection - Inspect for: Leakage at the surface of the flexible hose or its junction with the metal couplings. Blistering or abnormal deformation of the outer covering of the hydraulic or pneumatic hose. Leakage at threaded or clamped joints that cannot be eliminated by normal tightening or recommended procedures. Evidence of excessive abrasion or scrubbing on the outer surface of a hose, rigid tube, or fitting.	

Hydraulic and pneumatic pumps and motors - Loose bolts or fasteners Leaks at joints between sections Shaft seal leaks Unusual noise or vibration Loss of operating speed Excessive heating of the fluid Loss of pressure (flow test)	
Hydraulic and pneumatic valves - Cracks in valve housing. Improper return of spool to neutral position. Leaks at spools or joints. Sticking spools. Failure of relief valves to attain correct pressure setting. Relief valves checked as specified by manufacturer.	
Hydraulic and pneumatic cylinders - Drifting caused by fluid leaking across the piston. Rod seals leakage. Leaks at welded joints. Rod seals leakage. Leaks at welded joints. Scored, nicked, or dented cylinder rods. Dented case (barrel). Loose or deformed eyes or connecting joints.	
Hydraulic filters	
Wire ropes	
Hoist clutch lining	
Hoist drum brake bands	
Swing clutches (manufacturer specs).	
Master clutch (manufacturer specs).	
Open gears	
Boom stops	
'Electrocution Warning' sign visible from operators station.	
Lubrication	
Steering gears and connections	
Brakes (service and hand)	
Excessively worn or damaged tires.	
Exhaust systems.	
Back-up alarm.	
Air system.	
Suspension	

Electrical system.	
Mounting bolts	
Hydraulic hose and fittings	
Lubrication - Fluid levels. Crank case Battery Transmission Hydraulic fluids Auxiliary transmission Front axle Rear axle	
Overall Conditions - Glass Fire extinguisher Controls identified Load rating chart, operators manual, and maintenance manual. Hand signal chart Gauges Level Indicator Required controls Paint (touch-up for rust) Guards in place	
Operating test	
Inspected by:	
Crane#:	Date of Inspection:

Any deficiencies found that could reduce the crane's capacity or adversely affect its safety must be corrected before the crane is placed back in operation.

CHECKLIST # 5

ASSESSMENT CHECKLIST

USE OF AN OVERHEAD BRIDGE CRANE AS A WORK PLATFORM

This checklist outlines the generic safety assessment requirements to safely evaluate conditions to perform tasks utilizing an overhead bridge crane as a work platform. If unusual conditions are recognized or any question arises as to the safety of personnel, equipment and facilities, <u>STOP</u> until the questionable conditions are fully resolved. The supervisor in charge is responsible for reviewing and completing the safety checklist documentation.

Any question answered "no" shall be resolved before work is allowed to occur. If the crane is needed to be used to move a load, work from the bridge of the crane shall be suspended.

ASSESSMENT CHECKLIST

yes	no	
		Has work been evaluated and a crane determined to be the safest way to access the work location and perform work from the bridge platform?
		Has the assigned crane operator's training and qualification been verified?
		Are any support personnel (crane manufacturer, electricians, etc.) needed to assist in hazard evaluation?
		Have all required equipment changes/modifications (such as additional guarding, fall arrest attachment points, guardrails, etc.) as determined necessary by a hazard evaluation been completed?
		Have the requirements of OSHA Lockout/Tagout standard) been executed?
		Have appropriate tools, supplies, equipment, personnel protection and other safety equipment been obtained to safely perform work?
		Has the operator performed appropriate safety operational checks of the crane prior to allowing personnel to access the work platform?
		Has the work area on the crane been reviewed for hazards associated with live electrical parts, mechanical, or rotating parts on crane and contact striking hazards while moving the crane?
		Have the requirements of OSHA's Fall Protection standard been met?
		Has an operator/crew communication plan been established so that the crane operator will be notified before any work is performed from the crane?
		Does the crane operator know not to move the crane until it is determined that all employees on the crane are in location where they will not be exposed to injury?
		Has custody of operating controls been established?
		Has the floor level area under the work platform been properly isolated with barricades, flags, ropes, and/or warning signs?
		Have all administrative control measures/guidance been identified in an administrative document?
		Have employees been informed that no ladders are to be erected and/or used to gain access to areas that are not directly accessible from a crane without the use of a ladder?

yes	no	
		Are only "qualified persons" as that term defined at 29 CFR 1910.399, permitted to work on or near electrical equipment?
		Do all employees know that work from cranes shall be performed only when the crane is stationary?
		Are rail stops or other suitable methods used to prevent the crane from being struck whenever other cranes are on the same runway?
		Has a safe egress to and from the crane platform been provided?
		Has a pre-job briefing with all personnel involved in the work been conducted to review the hazards, requirements, and permits associated with this work?

CHECKLIST # 6

SUSPENDED PERSONNEL PLATFORM AUTHORIZATION FORM

Date:		Locale:	
The determination was made by _____ Title_____ that no other safe method is available and this is the least hazardous way to do this task.			
Person responsible for the hoisting operation:			
Platform Manufacturer	Model Number		Serial Number
Platform rated capacity for radius and configuration	Combined weight of loaded platform and rigging:		
Number of platform occupants	Approximate weight of occupants and equipment		Total lift weight B no more than 50% of rated capacity
A pre-lift inspection of all systems, controls and safety devices* was conducted by:			
Names of all employees attending a required pre-lift meeting and a trail lift performed immediately prior to placing personnel on the platform			

*Cranes and derricks must be equipped with (1) an anti-two blocking device that prevents contact between the load block and overhaul ball and the boom tip, or (2) a two-block damage-prevention feature that deactivates the hoisting action before damage occurs.

INDEX

Government Institutes Mini-Catalog

PC #	ENVIRONMENTAL TITLES	Pub Date	Price
629	ABCs of Environmental Regulation: Understanding the Fed Regs	1998	$49
627	ABCs of Environmental Science	1998	$39
672	Book of Lists for Regulated Hazardous Substances, 9th Edition	1999	$79
579	Brownfields Redevelopment	1998	$79
4100	CFR Chemical Lists on CD ROM, 1998 Edition	1997	$125
4089	Chemical Data for Workplace Sampling & Analysis, Single User Disk	1997	$125
512	Clean Water Handbook, 2nd Edition	1996	$89
581	EH&S Auditing Made Easy	1997	$79
673	E H & S CFR Training Requirements, 4th Edition	1999	$89
4082	EMMI-Envl Monitoring Methods Index for Windows-Network	1997	$537
4082	EMMI-Envl Monitoring Methods Index for Windows-Single User	1997	$179
525	Environmental Audits, 7th Edition	1996	$79
548	Environmental Engineering and Science: An Introduction	1997	$79
643	Environmental Guide to the Internet, 4rd Edition	1998	$59
650	Environmental Law Handbook, 15th Edition	1999	$89
353	Environmental Regulatory Glossary, 6th Edition	1993	$79
652	Environmental Statutes, 1999 Edition	1999	$79
4097	OSHA CFRs Made Easy (29 CFRs)/CD ROM	1998	$129
4102	1999 Title 21 Food & Drug CFRs on CD ROM-Single User	1999	$325
4099	Environmental Statutes on CD ROM for Windows-Single User	1999	$139
570	Environmentalism at the Crossroads	1995	$39
536	ESAs Made Easy	1996	$59
515	Industrial Environmental Management: A Practical Approach	1996	$79
510	ISO 14000: Understanding Environmental Standards	1996	$69
551	ISO 14001: An Executive Report	1996	$55
588	International Environmental Auditing	1998	$149
518	Lead Regulation Handbook	1996	$79
554	Property Rights: Understanding Government Takings	1997	$79
582	Recycling & Waste Mgmt Guide to the Internet	1997	$49
615	Risk Management Planning Handbook	1998	$89
603	Superfund Manual, 6th Edition	1997	$115
566	TSCA Handbook, 3rd Edition	1997	$95
534	Wetland Mitigation: Mitigation Banking and Other Strategies	1997	$75

PC #	SAFETY and HEALTH TITLES	Pub Date	Price
547	Construction Safety Handbook	1996	$79
553	Cumulative Trauma Disorders	1997	$59
663	Forklift Safety, 2nd Edition	1999	$69
539	Fundamentals of Occupational Safety & Health	1996	$49
612	HAZWOPER Incident Command	1998	$59
535	Making Sense of OSHA Compliance	1997	$59
589	Managing Fatigue in Transportation, *ATA Conference*	1997	$75
558	PPE Made Easy	1998	$79
598	Project Mgmt for E H & S Professionals	1997	$59
552	Safety & Health in Agriculture, Forestry and Fisheries	1997	$125
669	Safety & Health on the Internet, 4th Edition	1999	$59
597	Safety Is A People Business	1997	$49
668	Safety Made Easy, 2nd	1999	$59
590	Your Company Safety and Health Manual	1997	$79

Government Institutes

4 Research Place, Suite 200 • Rockville, MD 20850-3226
Tel. (301) 921-2323 • FAX (301) 921-0264
Email: giinfo@govinst.com • Internet: http://www.govinst.com

Please call our customer service department at (301) 921-2323 for a free publications catalog.

CFRs now available online. Call (301) 921-2355 for info.

Government Institutes Order Form

4 Research Place, Suite 200 • Rockville, MD 20850-3226
Tel (301) 921-2323 • Fax (301) 921-0264
Internet: http://www.govinst.com • E-mail: giinfo@govinst.com

4 EASY WAYS TO ORDER

1. Tel: **(301) 921-2323**
Have your credit card ready when you call.

2. Fax: **(301) 921-0264**
Fax this completed order form with your company
purchase order or credit card information.

3. Mail: **Government Institutes Division**
ABS Group Inc.
P.O. Box 846304
Dallas, TX 75284-6304   USA

Mail this completed order form with a check, company purchase
order, or credit card information.

4. Online: Visit http://www.govinst.com

PAYMENT OPTIONS

❑ **Check** *(payable in US dollars to **ABS Group Inc. Government Institutes Division**)*

❑ **Purchase Order** *(This order form must be attached to your company P.O. Note: All International orders must be prepaid.)*

❑ **Credit Card** ❑ VISA ❑ Master Card ❑ AMERICAN EXPRESS

Exp. ___ / ____

Credit Card No. _____

Signature _____

(Government Institutes' Federal I.D.# is 13-2695912)

CUSTOMER   INFORMATION

Ship To: (Please attach your purchase order)

Name _____

GI Account # *(7 digits on mailing label)* _____

Company/Institution _____

Address _____
        (Please supply street address for UPS shipping)

_____

City _____ State/Province _____

Zip/Postal Code _____ Country _____

Tel ( ) _____

Fax ( ) _____

Email Address _____

Bill To: (if different from ship-to address)

Name _____

Title/Position _____

Company/Institution _____

Address _____
        (Please supply street address for UPS shipping)

_____

City _____ State/Province _____

Zip/Postal Code _____ Country _____

Tel ( ) _____

Fax ( ) _____

Email Address _____

Qty.	Product Code	Title	Price

❑ **New Edition No Obligation Standing Order Program**
Please enroll me in this program for the products I have ordered.
Government Institutes will notify me of new editions by sending me an
invoice. I understand that there is no obligation to purchase the product.
This invoice is simply my reminder that a new edition has been released.

Subtotal _____
MD Residents add 5% Sales Tax _____
Shipping and Handling (see box below) _____
Total Payment Enclosed _____

15 DAY MONEY-BACK GUARANTEE
If you're not completely satisfied with any product, return it undamaged
within 15 days for a full and immediate refund on the price of the product.

Shipping and Handling	Sales Tax
Within U.S:	Maryland 5%
1-4 products: $6/product	Tennessee 6%
5 or more: $4/product	Texas 8.25%
Outside U.S:	Virginia 4.5%
Add $15 for each item (Global)	

SOURCE CODE: BP01